Out of the Cave

Out of the Cave

A Natural Philosophy of Mind and Knowing

Mark L. Johnson and Don M. Tucker

The MIT Press
Cambridge, Massachusetts
London, England

The MIT Press would like to thank the anonymous peer reviewers who provided comments on drafts of this book. The generous work of academic experts is essential for establishing the authority and quality of our publications. We acknowledge with gratitude the contributions of these otherwise uncredited readers.

This book was set in Stone by Westchester Publishing Services, Danbury, CT. Printed and bound in the United States of America.

Library of Congress Cataloging-in-Publication Data

Names: Johnson, Mark, 1949- author. | Tucker, Don M., author.
Title: Out of the cave : a natural philosophy of mind and knowing /
    Mark L. Johnson and Don M. Tucker.
Description: Cambridge, Massachusetts : The MIT Press, [2021] |
    Includes bibliographical references and index.
Identifiers: LCCN 2020048448 | ISBN 9780262046213 (hardcover)
Subjects: LCSH: Philosophy of mind. | Knowledge, Theory of.
Classification: LCC BD418.3 .J65 2021 | DDC 128/.2--dc23
LC record available at https://lccn.loc.gov/2020048448

10 9 8 7 6 5 4 3 2 1

Contents

Preface and Acknowledgments

This book is the result of several years of conversation and collaboration between a philosopher with pragmatist sympathies (Mark Johnson) and a psychologist and neuroscientist (Don Tucker). Don initiated the dialogue by suggesting that some of his experimental research supported the general philosophical and linguistic perspective emerging in Mark's work on how our bodies shape mind, thought, and language in a deep and pervasive way. We shared the conviction that it was now possible to craft a natural philosophy (i.e., an empirically grounded theoretical and practical explanation) of mind and thought that could explain how people actually process meaning and how they understand their world through their intimate visceral bodily engagement with their surroundings.

We were attracted by the possibility of explaining human nature, as well as our individual selfhood, as the result of two processes: (1) our ongoing evolutionary history that provides the architecture of our bodies and brains and (2) our individual cognitive and affective development over the course of our lives, which sculpts our neural networks. This evolutionary-developmental approach is at the core of modern theoretical biology, and it provides the basis for a natural philosophy that explains how brain architecture and neural connectivity give rise to mind, conceptualization, and reasoning.

This new natural philosophy of mind was not made from whole cloth, as if it sprang ex nihilo from a conceptual analysis of cognition. Its origin instead was a cobbling together of insights from biological psychology, evolutionary theory, neural modeling, research on cognitive predictions, affective neuroscience of emotions, the science of animal motivation, and research on how values shape meaning, thought, and knowing. No single method or body of experimental research was sufficient. What was required was a search for convergent evidence arising from multiple approaches and perspectives.

As we discussed what such a natural philosophy of mind would involve, it became clear that it would require a radical rethinking of certain traditional views about human cognition and behavior. We began to explore the implications of this new perspective on mind for the nature of consciousness, thought, meaning, language, knowledge, basic values, and the exciting yet troubling aspects of the burgeoning Age of Information in which we find ourselves. Realizing that we couldn't adequately manage all of this in a single book, we decided to focus on the fundamental questions of what *mind* is and how it is possible to *know* anything. That alone was plenty to keep us occupied, but it became obvious that you cannot pursue these matters without also saying something about all of the other issues we had been discussing. So, along the way, we have taken up topics such as the nature of selfhood, consciousness, motivation, information, meaning, abstraction, concepts, and more.

As an example of the radical implications of embodied mind, you cannot talk about mind without talking about selfhood. The *self* as conceived in this evolutionary and developmental framework could not be a fixed and static entity possessed of unique powers of thought. Indeed, the self cannot be a *thing* or entity at all. Instead, it emerges as a pattern of ongoing processes of organism-environment interaction, processes that are at once biological, interpersonal, and cultural. The self is always in process, and it changes with each activity of inquiry and knowing that it unleashes on its environment. As we shared our research and discussed various problems we were struggling with, we glimpsed the possibility of synthesizing several bodies of recent philosophy, psychology, and neuroscience into a view of human mind that was entirely nondualistic, embodied, and rooted in biological and social values.

Moreover, if the self is always in process, continually attempting to accommodate changing conditions and events in experience, then knowing is a motivated activity—an ongoing process—shaped by our deepest biological and cultural values, geared toward survival and enhanced well-being. Knowing cannot be an internal re-presentation of a prior fixed and finished reality, but rather is a means of reconstructing experience in adaptation to changing conditions. This makes knowing an exploratory, transformative, and projective process that must remain fallible and subject to subsequent correction, in light of newly emerging circumstances. Knowing is thus not a static relation of mind and world, but rather a process that

1 Toward a Natural Philosophy of Mind

Mind is fundamentally embodied. The beauty of the body and its world creating the mind is wondrous to behold. Everything we experience, know, feel, value, and do is the result of bodily processes of which we are seldom aware, but without which we could neither survive nor have meaningful lives.

As recently as thirty years ago, this was a radical claim. Human cognition was too often studied without any serious understanding of how our bodies shape both *what* and *how* we think, learn, and know. Today, scientific research supports and elaborates the view that mind and thought are embodied social processes. In this book, we attempt to summarize some of the exciting recent research that reveals how all of our higher-level cognitive activities are rooted in our bodies through processes of perception, motive control of action, and feeling.

We propose that a naturalistic theory of mind may unify the story of the human condition provided by the sciences and the humanities. When it is fully formed, this theory will explain how self-consciousness, conceptualization, language, reasoning, and knowledge could arise out of, and operate through, many of the same bodily processes we share with other mammals. The most complex levels of human cognition are organized through basic mammalian patterns of organism-environment interactions, combined in powerful abstractions within human bodies and brains. There are multiple levels of functional organization that make up what we call "mind." Consequently, we need multiple levels of inquiry (e.g., biology, psychology, phenomenological description, neuroscience, and neural network modeling) to understand what goes on at each level of increased complexity and functional emergence.

The resulting natural philosophy of mind situates all cognitive activities in ongoing, value-laden transactions with our biological, interpersonal, and cultural environments. Important insight comes from seeing how these highly complex functions are possible because they appropriate patterns and processes of our sensory, motor, and affective operations to construct abstract conceptualization and reasoning. The emerging neurobiological evidence, when organized within the perspective of a natural philosophy, provides a twenty-first-century answer to the ancient injunction to "Know Thyself."

From biology we understand how the process of mind in our daily reflections is the product of a long history of mammalian evolution that has generated the specific capacities for perception, bodily movement, emotions, feelings, thought, imagination, and language. Mind is not fixed and finished, but rather an evolving biological, cultural, and technological process. As we learn new things through our activity in the world, our brains are continually rewired in the ongoing development of experience. Each new conceptual organization is a new subjective understanding with implications for personal identity. Consequently, changes in knowledge are changes in who we are.

For many centuries, philosophers have taken mind and intelligence to be unique to humans, though perhaps granting limited mind-constituting capacities to certain "higher" animals (e.g., dogs, cats, dolphins, horses, ravens). However, with the arrival of artificial intelligence, our conceptions of mind now face new and intriguing challenges. Intelligent systems based on neural models can perform many of the cognitive functions we previously took to be exclusive to humans, and, in many cases, these computational systems outperform us mere mortals. In addition, the new field of *extended mind* research argues that mind is not locked up in individual brains and bodies, but instead extends beyond the confines of skin and skull out into informational structures in our environment, such as when we off-load memory, computation, and situational awareness onto our phones and tablets.

These rapidly changing developments in intelligent systems are at once marvelous and scary. In our hectic, information-flooded daily existence, some people become anxious over the way informational technologies seem to be taking over their lives. It is not just that we cannot keep up with new devices, but rather that these technologies define and control our values, goals, practices, and modes of communication without our knowledge or permission. Many scientists anticipate the day when computers will become conscious agents capable of self-directed activity that rival humans for control of their environment.

As it merges with biologically based cognitive science, the new computational science requires us to rethink many of our most deeply held assumptions. The concept of human nature is at stake here. We need to understand the implications of this artificial intelligence revolution for our grasp of the nature, purpose, and value of biologically and socially embodied human knowledge.

1.1   The Folk Theory of Disembodied Knowing

Given the importance of knowing for understanding our place in the world, it may seem surprising that hardly anyone can give an adequate account of the knowing process. Most people have only the barest clues to how mind and thought work, and this is true even for many who have taken courses in philosophy, psychology, and neuroscience. Mostly, we pick up bits and pieces about the nature of mind and knowledge that have been passed down over the last several generations. These unstated assumptions have become a widely accepted *folk theory*, which assumes that we have knowledge of something when we form ideas in our mind that correspond to how things are in our world. Acts of knowing are taken to be purely intellectual operations, rather than bodily processes. Over the past 2,700 years, philosophers and, more recently, psychologists and other cognitive scientists have developed more sophisticated versions of this simplified model of knowing. Yet, the core tenets have persisted over history and constitute what might be called the *folk theory of disembodied knowing*:

- From birth, humans begin to acquire skills for carrying out practical tasks (*procedural* knowledge or knowing *how* to do something), and they later develop capacities for theoretical understanding of their world (declarative knowledge or *knowing that* something is the case).
- *Knowing how* is geared toward changing the world through your actions, whereas *knowing that* provides an objective, impartial understanding of what exists, why things are the way they are, and why they behave as they do.
- Theoretical knowledge is believed to be the product of rational processes that represent, or mirror, reality.

- These rational operations are taken to be fundamentally different from, and independent of, perception and physical engagement in the world.
- Rational theoretical knowledge is independent of emotions, feelings, or values. It is assumed that allowing values to influence our knowledge processes would undermine the possibility of objective knowledge and truth.
- The mind is thought to be fixed and constituted, with all its capacities and structures intact, prior to any act of knowing. Knowing is either the addition of new mental content or else it is the discernment of new relations in existing content.
- The mind (or our "knowing self") consists of a set of cognitive faculties (such as perception, imagination, understanding, and reason) that take in sensory impressions and organize them into conceptual structures and sentence-like propositions concerning states of affairs in the objective, mind-independent world.
- We have knowledge whenever we have justified true beliefs concerning how things are in the world.

This view of knowledge will seem to most Westerners very intuitive, even self-evident. But that is only because we have inherited a host of deeply rooted cultural assumptions about mind, thought, knowledge, and values that give rise to a view of mind as disembodied. These assumptions are embedded in our language and our educational institutions, and are found even in our mainstream scientific, political, economic, religious, and philosophical systems.

1.2   Out of the Cave

Plato's famous Allegory of the Cave describes a metaphorical journey from ignorance and unjustified opinion all the way to a grasp of the ultimate, unchanging essences of things, which constitutes genuine knowledge. We humans are prisoners chained in the cave, seeing only shadows of objects on the cave wall. In Plato's version, the journey from darkness to light does not end when one of the cave dwellers in ignorance is freed from his shackles and emerges, blinded, into the light and fresh air of the natural world beyond the cave's entrance. For, on Plato's view, even the natural world, as perceived, is taken to be too changing and unstable to be the object of knowledge. Instead, says Plato, full knowledge requires transcending the changing events of the natural world perceived by our senses in order to grasp the pure, unchanging forms of what really and truly is, discoverable only in the intelligible world beyond all sense experience. Plato's allegory thus finds true knowledge only in what he calls the intelligible world of pure ideas that transcend the senses. In sharp contrast, we will be constructing a radically different conception of what it means to be "out of the cave." It is not rising above the natural world of sense experience. Neither is it a turning toward the sun, conceived by Plato as the metaphorical source of knowledge of what things are and how they interact.

Instead, being out of the cave means immersing ourselves more fully in the natural world, utilizing our embodied intelligence to survive and flourish. Knowing is a completely natural process, rooted entirely in our ongoing engagement with our physical and social worlds. There are no eternal forms or unchanging essences. What some philosophers have called an essence—a set of properties that allegedly uniquely defines the nature of an object or event—is merely a recurrent pattern of organism-environment interactions that people have found it useful to note for some purpose that matters to them. Essences of this sort *are* useful, but there is nothing absolute, universal, or eternal about them. We know in and through our embodied activities in our material, interpersonal, and cultural worlds. Our animality is the very condition of our knowing engagement with our world, rather than an impediment to knowledge.

We agree with Plato that the shadows and images on the cave represent a very partial, perspectival, and sometimes even misleading take on the nature of things. But this should be no cause for skepticism or despair, since *all* our knowledge is perspectival and partial. Consequently, our perception and action are not inferior modes of knowing, but rather the very means for our exploration of our world. The real illusion is that the mind is of different stuff than the physical world, and in this sense, Plato's error was to regard the shadows on the cave wall and the physical objects and natural events of the perceived world as inferior modes of understanding. We will argue that knowledge is of nature's objects, events, and processes—a completely embodied immersion in and engagement with our world. When we exit the cave, we come to see that the light of scientific evidence shows us how the human mind emerges from the very stuff of physical reality, in the biological process of evolution and in the processes of individual development. Exiting the cave requires us, first, to give up the myth of pristine objectivity and absolute knowledge, and, second, to launch ourselves out into our natural world in ways that are attuned to what our world affords us by way of survival, flourishing, and pursuit of well-being. Getting out of the cave is learning our place in the natural world.

1.3   The Embodied Mind Perspective

We review research in biology, cognitive science, neuroscience, and computational neural modeling to show how human minds are embodied in the deepest possible way. It is not only perception, feeling, and bodily movement that are rooted in structures and processes in our bodies and brains. The same holds for conceptualization and reasoning, which are also grounded in and shaped by our embodiment. Mind is an emergent functional organization of body-based processes.

The same sensory, motor, and affective processes underlying perception and action are recruited for so-called cognitive activities of conceptualization and reasoning. As we will argue in the pages that follow, abstract thought is not a transcendence of the body, but rather is inherently the result of body-based meaning making. All of the affective and cognitive operations we perform—from simple perception to our most impressive intellectual and artistic achievements—are affairs of the embodied mind. *What* we think and *how* we think depends on our brains and bodies as they operate in our physical, social, and cultural environments.

Our bodies provide our primary animate situatedness in the world (Sheets-Johnstone 1999). Out of organism-environment interaction arises the meaning we make of experience and all of the reason of which we are capable. Grasping the meaning of our surroundings makes it possible to survive, grow, move forward, instigate actions, and coordinate with other creatures in joint cooperative activities. Knowing is our way of trying to find our place in our surroundings. We will present evidence that knowing is based on expectancies developed over the course of our experience, which are then evaluated in relation to present experiences in our perceptual and motor interface with our world. We readily know what we can predict.

But when our lived experience fails to meet our preestablished expectancies, we have to recalibrate those expectancies so they are more in line with the actual course of experience. This is a different, more critically based mode of inquiry and knowing.

Most of the time, both our expectancies and our recalibrations happen automatically beneath the level of our conscious awareness. However, with the emergence of abstract thinking and language use, this process of self-adjustment in light of new experiences can sometimes be brought to reflective awareness, thereby becoming subject to conscious articulation and control. The very possibility of science depends on this capacity to be reflectively aware of this expectancy-testing-adjustment operation.

Knowing is about developing a suitably rich and deep understanding of the meaning of your surroundings (physical, social, and cultural) that enables you to function more or less successfully. When we find ourselves stuck in a situation because our habits and familiar expectancies fail to manage the current conditions, the challenge is to make a fresh inquiry into ways we might adjust our expectancies and habits in order to restore our effective agency in the world. Knowing is then a *doing*—an active transformation of our experience from a condition of indeterminateness, uncertainty, and confusion to a condition of restored fluid activity necessary for us to function well within our world. This conception of knowing as intelligent doing was set out by the pragmatist philosophers in the late nineteenth and early twentieth centuries. We have adopted John Dewey's brand of pragmatist philosophy for its merits in framing an appropriate philosophical perspective on knowing that is consonant with our current science. As we study the brain's mechanisms for prediction and testing the evidence of the world, we will see how the sources of knowledge are profoundly embodied, personal, practical, and activity oriented. This bodily basis remains the foundation, no matter how abstract or esoteric our knowledge might be. Whether in mathematics, logic, science, or the arts, the process of mind is an embodied, value-based process.

1.4 Meaningful Mind Science

Of all the marvelous expressions of mind, perhaps the most important is *knowing*. Our goal is to clarify the everyday process of knowing—about ourselves and the world—by understanding how this process emerges from the brain's biological workings, its adaptive mechanisms. Our goal is an account of mind and knowing that is both scientifically supported and existentially meaningful—an account that helps us to understand what it means to be human and that bears directly on how we ought to live. Making a science of the mind meaningful for subjective experience is a challenge. It is one that has not been addressed very well by either philosophy or psychology, and perhaps even less so by neuroscience. Nonetheless, we think that discovering the nature and meaning of the process of knowing should be among the most important challenges for philosophers and psychologists. What follows are some key desiderata for a meaningful science of mind.

1. *Meaningful mind science must appreciate and explain the role of subjective experience in knowing.*

In science generally, and in psychology specifically, subjectivity has been avoided, as if admitting it would taint the scientific method. This avoidance reflects an error in our understanding of what a scientific psychology ought to involve.

Scientific knowledge can never be value neutral, but it can seek validation within knowledge communities (e.g., scientific disciplines, artistic communities, philosophical traditions, and technologies) and transcend limitations of personal assumptions, values, or interests.

Subjective experience weaves feelings, perceptions, motives, actions, and reasoning into the process of knowing. An important requirement for an adequate scientific theory of the human mind is a full account of all of its causal influences, including the urges, motives, and desires that are integral to personal experience. We will see that in order to frame a theory of mind and knowing with biological principles, we must understand how the brain's cognitive capacities—for attention, memory, and planning—are only possible because they are regulated by motive controls. These are the neural control systems that evolved to direct behavior for the essential tasks of survival and reproduction. They generate the emotional qualities and motive directives of subjective experience, and they drive learning. Motive controls give direction and meaning to our thinking and knowing.

2. *Meaningful mind science must give an account of the whole person who knows.*

First-generation cognitive science was relatively disembodied and focused primarily on cognition (e.g., concepts, propositions, logic, information processing, and artificial intelligence; Varela, Thompson, & Rosch 1991). Second-generation *embodied* cognitive science and neuroscience moves toward an appreciation of the entire human being, based on a deep understanding of embodied subjectivity (Feldman 2006; Lakoff & Johnson 1999).

A similar broadening of horizons has occurred with the emergence of embodied cognition theory and cognitive linguistics (Feldman 2006; Johnson 2007; Lakoff 1987; Langacker 1987–1991, 2002). As cognitive neuroscience has increased recognition of the importance of the brain and the biological basis of human nature, embodied cognition theory has provided a deeper appreciation of the phenomenology of our bodily engagement with the world and increased recognition of the key role of emotions and feelings in meaning, thought, and language (Damasio 1994, 1999, 2010). This approach leads to a more holistic understanding of the human experiences incorporated in the mind. Mind is the whole organism in interactions with its environment, rather than a narrow vehicle for rational thought (Gallagher 2005; Lakoff & Johnson 1999).

3. Science and philosophy must coevolve.

In her groundbreaking book *Neurophilosophy* (1989), Patricia Churchland optimistically looked forward to a creative coevolution of philosophy and cognitive neuroscience. These fields are finally catching up to her vision. A broader philosophical analysis may improve the science of the mind by making it more self-critical, while advances in cognitive neuroscience allow us to place the philosophy of mind on a more rigorous scientific basis. Our biological functions give rise to our subjective experience of self, thought, and knowing. Principles of biology explain key aspects of the mind's processes and structures in relation to the brain's function. These same principles lend insight into how we process personal experience, meaning, and behavior. A principal goal of this book is to understand the human brain in the context of the biological and social processes that generate and shape our subjectivity.

4. Meaningful mind science must recognize the central role of values and feelings in knowing.

One of the more important discoveries of a biologically based cognitive science is the central role of values in determining *what* we know and *how* we know it (Damasio 1999, 2003; Tucker 2007; Tucker & Luu 2012). Knowledge depends on our ability to evaluate evidence to learn whether it confirms or denies our expectancies and hypotheses. Mid-twentieth-century philosophy of science has shown there is no such thing as complete and final confirmation of a theory, though some theories can be rejected on the basis of falsifying evidence (Kuhn 1962; Popper 1959).

Yet, evidence isn't enough. We can gather evidence in the form of measurements, images, or statistics, but this may remain just raw data, not information, if it fails to have meaning in the context of the questions we bring. To be significant, the process of knowing must be grounded in meaning.

From a philosophical perspective, the meaning of something is grounded in the experience it evokes. This meaning helps to interpret past and present experiences, and what it portends for future (possible) experiences. Meaning is relational—it involves relations among experiences *for a person*. Therefore, knowledge is about both *relations in the world* and a *person's relations to that world*. To have meaning, information must have personal value. Dry, irrelevant data isn't meaningful. It remains as data. Instead, the most meaningful things are the valued properties of experience, the things about which we care deeply. To be meaningful, knowing must therefore be grounded in our values.

Values have an important basis in emotions and feelings (Damasio 2003, 2018). As we will see in the following chapters, today's cognitive neuroscience is teaching us to find the elementary basis of value judgments in the brain's biological roots, the motive regulatory systems. These motive systems give rise to familiar but complex phenomenological (experiential) patterns in our emotions and mood states. They engage integral regulatory controls provided by the brain's motivational circuits, including the visceral controls from the limbic system and the chemical neuromodulator systems of the brain stem.

There is a growing body of empirical research showing that gaining knowledge depends as much on motivation as on native intelligence. To understand how information gains meaning, which is crucial to the process of knowing, we must therefore appreciate the emotional and motivational controls that guide cognition. These controls are not only integral to effective thinking, but they become organized and abstracted in complex ways in the conceptual structures of our values.

5. *Meaningful mind science must appreciate the psychological experience of information in reducing indeterminacy and anxiety.*

Another critical dimension arises when we process information. The root of the term "information" is *in-form*, meaning to change the form of the mind. We can begin with the technical definition of information in computer science. In Claude Shannon's mathematical theory of communication (Shannon and Weaver 1949), the *information* of a message is measured as the *reduction of the uncertainty* of the receiver (the person being informed). The uncertainty can be described objectively in mathematical terms (bits) in a particular communication context. To achieve this objective description, it is important to recognize that the context of the communication must be fully specified (by the omniscient observer). Within the idealization of a fully known context, the value of the specific piece of information—in reducing the receiver's uncertainty—can then be known and quantified.

There is a subtle yet profound implication of this objective formulation of information that may not be obvious in our daily use and that we will consider carefully at several points in this book. The communication context—all the facts and specifics related together—is essential in determining the meaning of information. Information is therefore *relational* in two important senses: (1) the value of a given piece of information depends on its relation to the larger context and (2) information is valuable only relative to particular agents, that is, knowers and actors. This makes it difficult, or even impossible, to give an objective definition of a specific piece of information in isolation from its context and relations between sender and receiver. Instead, meaning is contextual and therefore inherently holistic, situational, and complex. Meaning is grounded in personal values, not just in objective data.

Uncertainty is both a logical and a psychological issue. Indeterminacy and uncertainty often unfold from the ambiguous relationship between an event and its larger context. However, the psychological state of uncertainty also depends on the person who is in that state—their needs, values, and interests.

We will separate two quite different determinants of the meaning of information and the process of knowing. One is objective: the relation of information to the evidence of the larger experiential context. The other is subjective: the relation of information to personal values. Both of these determinants of information are measured by, and also influenced by, our subjective degree of certainty. In the chapters that follow, we will see how the organization of cognition in the human brain is grounded in both domains: the accuracy of contact with the world and the meaning for personal values. The biology of cognition consolidates experience through negotiating between these two boundaries of meaning, each represented in unique domains of the neural networks of the cerebral hemisphere, one domain close to the mind's visceral core, the other aligned with the sensory and motor contact with the world. Even for abstract concepts, it will be instructive to consider how a theory of mind must include these dual domains of organizing knowledge.

6. *Meaningful mind science must recognize the key role of culture in organizing knowledge.*

Even as we emphasize the biological basis of mind, it is clear that we also need to consider the role of social interaction and culture. The evolution of human intelligence has occurred through the socialization made possible by an effective family, social, and cultural context. It was this supportive social context that allowed human children to continue their juvenile neural plasticity, and the resulting learning capacity, well into adulthood, thereby developing the high-level cognitive functioning that is most distinctive of our species. In this extended learning process, it is important to appreciate how the mind becomes more complex and more fully differentiated as the child's educational and cultural experience continues over a long childhood.

In humans, the mammalian dependency on the parent has been greatly extended, well into adulthood. Our cultures literally shape the minds that develop within each unique family, community, and society. The increasing complexity of our modern culture is, in fact, a key issue that sets the stage for this book. The last several decades have seen a remarkable advance in the information technologies that are now integral to our daily lives. These information technologies are defining a new cultural context for the evolution of mind that we need to understand. If we can employ a meaningful mind science, we might gain new perspective on the new problems of our changing society.

1.5   The Cautionary Tale of the Two Cultures of the Academy

Integrating scientific insight with personal meaning has not been so easy. In fact, certain people, especially scientists, become highly trained in dealing with evidence but often have limited interest or insight into human values and motivation. Others, such as humanists, become most concerned with the subtleties of human values, even if they do not have the training or discipline for understanding the scientific study of the brain and its intelligence.

The intellectuals of modern societies, including teachers and professors in our colleges and universities, seem to gravitate to one intellectual skill at the expense of the other. In remarking on the English educational system of the 1950s, C. P. Snow described the sciences and the humanities as the two cultures of the academy (Snow 1959). He criticized his humanist colleagues for failing to understand the most basic scientific discoveries, even as he criticized the scientists' ignorance of the world's literary understanding of human values and cultures.

In the study of human nature, scientists and humanists often find themselves on opposite sides of a deep divide in what stands for knowledge. As we have begun our analysis of the process of knowing in terms of the dual skills of understanding evidence and understanding values, it seems clear that these are not easily developed together.

A lesson from the continued separation of the two cultures in the modern era may be that we naturally gravitate to different relationships with knowledge. On the one hand, knowing is a personal, subjective relationship, where knowledge has felt meaning for understanding ourselves and who we think we are. This is the position of the humanities, where the emotional and value implications may be the most important feature of knowledge, leading knowledge to be relative to the person's or culture's values. Carried to the extreme, this emphasis leads to the absolute cultural relativism of much postmodern humanities discourse.

On the other hand, knowledge for the scientist is too often strictly a matter of evidence. Theory is of course necessary. Yet, the history of science shows that most scientists will be dragged into a new theoretical framework only by the crushing weight of irrefutable evidence. There is precious little insight into the subjective process of knowing and the implicit values that shape the mind's conscious products.

As we propose that the process of knowing requires both the understanding of evidence and the clear motivation from articulated values, we join Snow in rejecting the two cultures of the academy. Science must strive for objective knowledge, with skill in understanding the evidence, to be sure. Yet, when it is applied to the human mind, science should be expected to provide insight into the process of knowing, including what appears to our subjective perspective. Similarly, the humanities are concerned with the search for human meaning in the subjective aspect, with the complexities that accrue to individual minds in their unique cultural contexts. Yet, those humanists who are ignorant of the progress of the science of the mind risk missing crucial explanations of how the uniquely human mind has evolved and how individual development within a culture then shapes our capacities for meaning, valuing, thought, and language.

Despite the continuing separation of the two cultures, there are some promising recent developments that provide examples of a fruitful collaboration between the sciences (e.g., biology, neuroscience, and cognitive science) and humanistic disciplines (e.g., literary theory, ethics, and social theory) (Fauconnier & Turner 2002; Slingerland 2008; Turner 1991; Wehrs & Blake 2017). Works such as these show that a cooperative interaction is both possible and highly desirable.

1.6   The Plan of This Book

The natural philosophy of mind is emerging through the convergence of biology, psychology, computer science, and philosophy. Our goal with this book is not just to chronicle this emergence, but also to contribute to it, with original ideas that may help carry it forward. To start with a basis in philosophy for the general reader, chapter 2 will review classical questions about the nature of knowledge, along with a critique of the proposed answers that have shaped the collective Western intellectual tradition that we all share. Chapter 3 then takes on the key issue of the relation of the knower to what is known. Even though we want our knowledge to be objective, and not just a matter of opinion, an important theme in philosophy has been appreciating how the self is implicit in each person's process of mind. This role of the embodied, social self in forming concepts will be a key theme throughout the book.

A good philosophy is not just a collection of ideas, but a system, a coherent way of knowing. In chapter 4, we outline the philosophy of American Pragmatism (especially that of John Dewey) as a candidate for the best example of a complete philosophy that is fully compatible with scientific evidence and theory, on the one hand, and with the questions that arise from personal experience, on the other. Although it was most fully developed at the beginning of the twentieth century, Pragmatism still may offer a model for a philosophy that addresses the big questions while remaining open to the new insights of scientific inquiry.

As psychology developed as a science, you might think it would have started from the big questions that Pragmatism framed so clearly, such as how our abstract knowledge could emerge from our biological capacities, how our values come to guide our appraisals of important situations, or how we search for new knowledge based on what we already know. In chapter 5, as we frame some of the issues in the development of psychology as a science for the general reader, we find this presented quite a challenge to psychologists. To be scientific, as we will see, most research psychologists decided they had to start with more limited, behavior-based questions that presuppose a somewhat impoverished view of experience.

As a result, the academic psychology of the twentieth century was dominated by a fairly narrow approach, where subjective experience was seen as mostly irrelevant, and behaviorism denied mental life as a topic for science. Even as cognitive science emerged in the second half of the century, it did so by rejecting humanistic questions altogether. At about the same time, the "analytic" approach in philosophy attempted to reduce the questions of the basis for knowing to objective logic of propositions and semantic constructions. In both approaches to achieving pristine objectivity, subjective meaning was sacrificed.

In chapter 6, we consider some of the extensive evidence for the embodiment of mind, thought, and language that has emerged over the past four decades, within what is known as embodied cognition theory. Much of this evidence comes from cognitive linguistics, which investigates how meaning, thought, and language arise from sensory, motor, and affective processes. This body-based meaning is then recruited for abstract thought via conceptual metaphors that use structures and processes in a sensory or motor domain to structure our understanding of an abstract target domain. Conceptual metaphor thus turns out to be an essential process of human abstraction, while remaining firmly embodied. This primarily linguistic evidence for the embodied mind sets the context for the subsequent neuroscience accounts in the remaining chapters of how thought arises in neural tissues and bodily structures and processes.

To frame the biological basis for a natural philosophy of mind, chapter 7 then digs into the anatomy of the human brain. We explore the hypothesis that brain and body architecture supports the basic functions of mind. Mammalian evolution has so far resulted in the current functional organization of the human brain capable of activities of perception, bodily movement, action planning, inference, and more. Neuroanatomy reveals how the hierarchical organization of different types of sensory and motor cortex gives rise to perception and action. One of the most important discoveries is the central role of the limbic system in providing the motivational and value-shaped unifying context for these perceptual and motor processes. Basically, the sensory and motor networks of the neocortex regulate the traffic with the external world (in perception and in action), while the visceral networks of the limbic cortex regulate the brain's traffic with the internal milieu of bodily needs, such as the maintenance of homeostasis. This suggests that all cognition is driven by our deepest motivational processes and our deepest biological and social values.

To consider the motivated, biological process of mind in more detail, chapter 8 then reviews current ideas on how motive controls regulate memory and cognition. One novel implication is that the process of knowing cannot proceed without motive control. The separation of cognition from motivation and emotion turns out to be an academic pretense—something that philosophers and psychologists might wish to be true but biologists are learning is not possible. The positive implication is that we can explain the role of values in organizing experience, with a new explicitness that is refreshing as well as instructive for an account of situated, value-based modes of knowing.

In order to align the motive control of cognition and memory with the anatomy we studied in chapter 7, the analysis in chapter 8 draws from current theory in computational neuroscience to consider how experience shapes the vast, constantly changing, neural connectivity among functional brain regions. This research supports the theory of *predictive coding* to suggest how the process of mind is implemented across the linked networks of the limbic system and neocortex. One significant implication is that all our knowing is motivated by our biological and social values. Another is that knowing involves the projection of expected events, which are then assessed in relation to actual experiences and recalibrated when there is a discrepancy between predictions and actual events. This is the deepest, most fundamental basis for inquiry and knowing.

Chapter 9 delves more deeply into the motive control of bodily behaviors and how that motivational structure is appropriated for higher cognitive functions, such as conceptualization and reasoning leading to knowing. These deep motivational processes underlie our acts of knowing, both as unconscious regulatory processes and in our conscious reflective modes of inquiry. These ideas follow directly from the functional neuroanatomy of chapter 7 and the cognitive and computational neurophysiology of chapter 8, and they help us become more aware of the roots of our thinking and knowing in emotions, feelings, values, and motivational systems. The motive controls that we found to regulate bodily behavior also structure our conceptual thinking. We think by feeling. When we feel elated, we can think expansively and then predict the big picture of the impending future; when we feel anxious and uncertain, we can focus and then exercise caution in critical thought. Although these are novel ideas, they follow directly from the anatomy and biology that we review, which provides new insights into the phenomenology of personal experience.

Chapter 10 explores what concepts become, once we realize that they are an essential part of the basic motivational, value-driven control of thought and action. Concepts are patterns of neural connectivity based on past experiences that establish expectancies concerning which experiences will be associated with particular objects, events, and actions. Concepts are thus predictions about what specific objects or events will afford us by way of experience. They depend on our values and interests, and they are tools for the maintenance and transformation of experience. Once we appreciate that concepts are both motivated and value dependent, we can investigate how motivational and affective processes operate, not just in our unconscious processes of perception and action, but also in our personal decisions and cognitive styles of thought. This explains why, in personality development, some folks manifest more impulsive and extraverted modes of self-regulation that foster a more impressionistic and holistic integration of conceptual structure, while others adopt more constrained and introverted modes of self-regulation that favor more analytic and focused forms of cognition, suggesting greater differentiation in conceptual structure and therefore greater analytic attention to detail.

Abstraction involves both hierarchical organization and unifying integration of different parts or dimensions of a thing. In chapter 11, we follow up Piaget's suggestion that abstraction requires some measure of self-awareness, which only develops fully in the second decade of a child's life. Abstraction requires that we stand back from the holistic identification of self and object experienced as an undifferentiated sense of elation. We have to suspend this blending of self and world temporarily and take up a more critical stance that allows for analytic discrimination, which is characteristic of the more anxious, self-critical mode of engagement. The stances of elation and anxiety both give rise to knowledge, but only the anxious mode makes a full-blown critical attitude possible. That mode of being in the world gives rise to a more empirically minded perspective, which, in turn, provides the basis for scientific modes of inquiry. However, science is no different from any other mode of experience and inquiry in its dependence on values and its rootedness in our motivational systems.

At the same time that a natural philosophy teaches us how to participate more consciously in the process of subjective experience, it also must explain the nature of objective reality and how the mind can grasp this nature in spite of our limited conceptual capacities and our implicit motive biases. Here, the scientific concepts of information and complexity provide general ways of understanding the structure of the world that must be approximated by the structure of mind. Chapter 11 reformulates the philosophy of knowing in terms of the complexity of reality, as that notion has been explored in information theory. Complexity in conceptual structure arises in response to the complexity of our natural environments. Concepts are not abstract entities existing in some mysterious mental space. Instead, they are patterns of neural connections that constitute expectancies about what particular objects and situations will mean to us. They are anticipations of experience. They are thus always in process and often have to be recalibrated in cases where our actual experience diverges from our projected expectations. Our critical ability to revise our conceptual expectations requires us to separate ourselves from our immediate immersion in our present experience—pulling back in self-awareness to assess the interface between our prior concepts and our ever-developing experience. We discover that what is required for truly abstract thought is not just consciousness but self-consciousness—the capacity of mind to direct its own cognitive process.

Chapter 12 takes stock of what this developing natural philosophy of embodied mind entails for our ability to know both ourselves and our world. We summarize our main claims about what a meaningful mind science involves, and about the nature of mind and knowing. Mind has both biological and sociocultural origins. It arises from the functional architecture of the brain and body, and from our social interactions with other people. Knowing is an entirely natural process, which depends on the functional architecture of the brain (and body) and on the individual development of each person over the course of their lifetime. The growth of knowledge is also a constructive transformation of the self, so that each new act of knowing creates a new self. Knowing is embodied, first, in the sense that it operates with sensory, motor, and affective processes that are mostly automatic and unconscious, and, second, in the sense that it is profoundly shaped by our motivations, interests, and values. All thought and knowing are shaped by our basic biological motor control systems, and knowing is therefore motivated by organism values such as homeostasis and allostasis that govern higher cognitive acts too. Our more reflective conceptual modes of thought repurpose those bodily operations for abstract thought. Human knowing is inescapably partial, tentative, and fallible, but it is also, when critically tuned, capable of giving us genuine knowledge that can aid our quest for survival, growth, well-being, and well-doing.

2 The Philosophical Quest for Ultimate Knowledge

2.1 The Need for a Theory of Embodied Knowing

Our goal is a scientific account of the development of the brain and body as the basis for knowledge, in both its objective dimension rooted in our engagement with the world and in its subjective dimension as meaningful for our lives. We address the role of the body in meaning, perception, experience, learning, conceptualization, self-awareness, selfhood, values, and anything connected with what is colloquially termed "mind." In later chapters we will explore how some of the structures and processes of the always-developing brain give rise to various modes of understanding and knowing. This account reveals that all our knowing is *embodied* (rooted in our bodies and brains), *embedded* (incorporating and interacting with our environment), *enactive* (continually generating experience and selfhood), and *value based* (grounded in our motivational and emotional systems and geared toward action in the world). Knowing of this sort is engaged, practical, existentially charged with meaning, and intimately tied up with our developing selfhood. It is a full-bodied, full-blooded exercise in learning the meaning of our world.

It is disconcerting to discover that philosophy, until quite recently, has had almost no engagement with the scientific research of the sort we describe in this book. Philosophy has long prided itself on being the discipline best equipped to tell us what knowledge consists in and how it works. Instead, what one too often finds are armchair analyses of our culturally learned concepts of meaning, knowledge, reference, and truth, but without any serious grounding in the cognitive sciences. Peruse any text or anthology that takes an Anglo-American analytic philosophy approach and you find there is an initial distinction drawn between kinds of knowledge (e.g., knowing that vs. knowing how vs. knowing by acquaintance). Usually, this is followed by an assertion that *knowing that* is the primary mode of objective knowledge of the world, so that the principal focus should be on how it is possible for inquirers to justify certain propositional claims as known and true.

As a representative example of this kind of focus of contemporary epistemology, consider the opening lines of the entry on the analysis of knowledge from the *Stanford Encyclopedia of Philosophy*:

> The project of analysing knowledge is to state conditions that are individually necessary and jointly sufficient for propositional knowledge, thoroughly answering the question, what does it take to know something? By "propositional knowledge", we mean knowledge of a proposition—for example, if Susan knows that Alyssa is a musician, she has knowledge of the proposition that Alyssa is a musician. Propositional knowledge should be distinguished from knowledge of "acquaintance", as obtains when Susan knows Alyssa. The relation between propositional knowledge and the knowledge at issue in other "knowledge" locutions in English, such as knowledge-where ("Susan knows where she is") and especially knowledge-how ("Susan knows how to ride a bicycle") is subject to some debate (see Stanley 2011 and his opponents discussed therein).
>
> The propositional knowledge that is the analysandum of the analysis of knowledge literature is paradigmatically expressed in English by sentences of the form "S knows that p", where "S" refers to the knowing subject, and "p" to the proposition that is known. A proposed analysis consists of a statement of the following form: S knows that p if and only if j, where j indicates the analysans: paradigmatically, a list of conditions that are individually necessary and jointly sufficient for S to have knowledge that p. (Ichikawa & Steup 2018)

This summary passage is a perfect example of the perspective, focus, and style of analytic philosophy treatments. It acknowledges other kinds of knowing, but focuses almost exclusively on descriptive (propositional) knowledge. Its central concern is how propositional knowledge claims can fit (or fail to fit) an objective state of affairs in the world. Finally, it says nothing about what knowledge means to us, and completely ignores the crucial subjective components of knowing that make knowledge relevant and important for our lives. An ordinary person, unschooled in the abstruse, abstract, and formal intricacies of contemporary analytic philosophy, might wonder what any of this has to do with knowing as it operates in our daily lives. How did philosophy come to narrow its scope and methods so much that it overlooks most of the body and brain processes that make knowing possible and meaningful? Why did it come to focus only on justifying one particular type of knowledge claim, namely, propositional assertions? How did philosophy of knowledge become primarily an "S-knows-that-p" episte-mology? How did philosophy become so alienated and detached from our ordinary practical lives, in which knowing is a doing—an activity—and a profoundly personal and emotional affair?

It behooves us to take a brief journey back through some key events in philosophical treatments of knowledge that have profoundly influenced how we think about knowledge, culminating in an almost exclusively conceptual and propositional focus. It would be hard to overstate the difference between the embodied notion of understanding and knowing that we propose in this book and traditional proposition-based accounts of knowledge. Propositions do sometimes play an important role in our knowledge practices, but, as we will argue later, these have to be understood and explained in relation to the kinds of embodied motivational processes that we will identify as central to all kinds of knowing, and not merely to propositions. In other words, language-centered knowledge practices actually depend on mostly unconscious, prereflective, and value-based embodied modes of knowing. So, instead of starting (and ending) only with propositions and linguistic phenomena, we argue for the need to start with body-based knowing and eventually build up to propositional knowledge, which is itself a form of embodied knowing.

To see how contemporary epistemology got off track, we need to exam-ine some important historical moments in philosophical treatments of knowledge that led the entire field of epistemology to ignore or downplay the crucial role of embodiment, motivation, emotions, values, and action in what and how we know.

We focus on four profoundly fateful moments in Western epistemology (Plato, Aristotle, Descartes, and Kant), which set the course for subsequent attempts to explain knowledge. We suggest that classical Greek philosophy left us with an ontology (a theory of being) according to which our world divides neatly into two fundamentally different and separate ontological realms. One is the temporally embedded and always changing perceptual realm of experience and bodily action. The other is an allegedly intelligible realm of eternal, unchanging, and fixed essences graspable by mind. The perceptual realm consists of objects and events causally interacting and changing over time. Because it is variable and changing, the perceptual realm cannot supply any fixed standard of knowledge, and it is therefore taken to be the basis of mere belief and opinion associated with practical affairs of living. The intelligible realm is regarded as transcending time and change, thereby providing the possibility of unchanging universal objective knowledge. The distinction between theoretical knowledge and practical knowledge fits well with the supposition of two different realms. Theoretical knowledge is an intellectual grasp of fixed and eternal essences, whereas practical knowledge focuses on the contingencies of everyday activities in which we seek to achieve our practical ends and goals. Practical modes of engaging the world are deemed epistemically inferior because they deal with contingencies and probabilities, rather than necessities and certainties.

2.2   The Metaphysics of Eternal, Unchanging Knowledge in a Precarious World

Although there are historical antecedents in pre-Socratic philosophy, we begin with Plato's and Aristotle's more elaborate and influential articulations of knowledge as rooted in a realm of unchanging essences. In *The Quest for Certainty* (1929/1984), John Dewey described, in a detailed, nuanced, and highly insightful manner, the general context in which classical Greek views of knowledge emerged and have, in various forms, been carried forward into the present day. He opens the book with an important general observation about how the human condition—precarious and fraught with peril—drives us to seek some means by which to manage the uncertainties and contingencies of life. "Man who lives in a world of hazards is compelled to seek for security" (3). One is reminded of the Hebrew lament, "Yet man is born unto trouble, as the sparks fly upward" (Job 5:7, King James Version). At birth, we are thrown into a world that continually challenges our ability to make sense of it and to control the forces that affect our lives for better and worse. No sooner do we get settled into habits of perceiving, thinking, and doing than new conditions emerge to unsettle those habits, requiring us to reconfigure ourselves or our surroundings if we hope to come back into relative harmony with our environment.

The early Greeks were painfully aware of life's perils and difficulties. They lived in a puzzling and confusing world subject to the whims of gods. They propitiated the gods with prayers and offerings in the hope that they would look kindly on them in the face of life's many uncertainties and difficulties.

Yet, even in the midst of such relentless peril and bewilderment, the earliest Greek nature philosophers, in the seventh and sixth centuries BCE (e.g., Thales, Anaximenes, Anaximander, Parmenides, Heraclitus), claimed to perceive an underlying order in nature (*phusis*)—an order discernible to some degree by human intelligence. If this is so, then perhaps humans are not completely at the mercy of forces beyond their control. This recognition of a rational order (*logos*) underlying natural events and processes in the cosmos is surely one of the most important discoveries in human history. Insofar as we can predict certain events, by discerning why things happen the way they do (the *logos* underlying natural happenings), we can also intervene to influence the course of events. We are no longer helpless in the face of life's contingencies. Instead, we have at least some measure of control over what happens and can arrange our lives adaptively in relation to the perceived orderliness of nature.

The discovery of various skills and arts (*technés*) for managing the affairs of daily living was evidence that nature manifested a rational order, at least in certain domains of experience. Martha Nussbaum explains the Greek conception of *techné* as our greatest rational resource for managing *tuché* (chance):

> *Techné*, then, is the deliberate application of human intelligence to some part of the world, yielding some control over *tuché*; it is concerned with the management of need and with prediction and control concerning future contingencies. The person who lives by *techné* does not come to each new experience without foresight or resource. He possesses some sort of systematic grasp, some way of ordering the subject matter that will take him to the new situation well-prepared, removed from blind dependence on what happens. (Nussbaum 2001, 95)

The skills and practices of farming, home building, tool production, and medicine are instances of intelligence operating to make our lives better and less subject to contingencies.

From a practical perspective, you might conclude that the development of various technologies represents the only knowing we need for the maintenance of our affairs. Unfortunately, early Greek philosophers were not satisfied with know-how as a model of human knowledge. *Techné* applied only to contingent natural processes governing things that come into being, change over time, and eventually pass out of existence. However, early Greek nature philosophers typically sought to discover a truly universal basis for what they thought was knowledge in the eminent sense, that is, the grasp of fixed essences and their relations. In this way, the Greeks introduced a fateful distinction that has haunted Western philosophy, namely, a distinction between two worlds or realms of Being—the ever-changing contingent world of objects and events grasped through bodily perception, and the fixed and timeless world graspable by the intellect (*nous*).

The dualistic logic and corresponding ontology took the following form. True (genuine) knowledge must be of what is, of that which is for all time, in all places, eternal and complete. Such knowledge is the only basis that allows us to rise above the vicissitudes of life and the partiality of our perspective on the changing world. Objective universal knowledge must be based on understanding the essences of things that make them what they are and govern their interactions in the world. There thus arose a primary dualism between fixed, eternal objects of the intelligible realm versus the changing objects of the perceptual realm. About such transitory objects, we cannot have certain knowledge, only belief (opinion). Consequently, the Greeks concluded that the objects of intelligence or reason were eternal, unchanging forms or essences, whereas the objects of the practical arts (*technés*) were the changing perceptible materials and instrumentalities of ordinary life.

Early Greek mathematical thinking provides an excellent illustration of this bifurcation of reality, so as to secure absolute knowledge. Pythagoras, known today mostly for his geometry, recognized that perceivable objects could be characterized mathematically by numerical proportions and relations. Objects were seen to be both measurable and governed in their interactions by numerical relations. This was a discovery of monumental proportions because it supposedly revealed a rational mathematical order behind the objects of perception. Hence, the Pythagoreans created an ontology based on their overarching metaphor, THE REAL IS NUMBER[1] though they regarded this as a literal truth, rather than a grounding metaphor. That is, the being of any

---

1. Throughout this book we will be making frequent reference to fundamental conceptual metaphors that structure most of our abstract conceptualization and reasoning. In their book *Metaphors We Live By* (1980), Lakoff and Johnson established the standard format for indicating the names of conceptual metaphors. Those names are given in SMALL CAPS, with the initial letter in each word of the name given in Large Caps. As we explain later, it is critical to keep in mind that the metaphors are not their names, nor are they linguistic as such. Rather, they are conceptual and consist of a cross-domain mapping. We use the names as a convenient way of indicating the source and target domains that make up the metaphor. Lakoff and Johnson later used the same SMALL CAPS indication for the names of embodied structures of meaning they called image schemas. In chapter 6 we provide a detailed account of the nature and workings of conceptual metaphor that is the basis for our brief discussion of key metaphors in the history of Western philosophy over the next four chapters.

particular kind of thing was determined by the mathematical properties and relations that defined it. For example, the so-called Pythagorean theorem (namely, that for any right triangle, the square of the hypotenuse is equal to the sum of the squares of the other two sides) could be materially demonstrated, in a sense, by drawing squares on each of the three sides of a right triangle and then cutting up the squares of the two shorter sides and showing that they exactly fit into the square drawn on the hypotenuse. Hence, the Pythagoreans concluded that reality ultimately has a mathematical character, and physical objects are constituted by their mathematical properties and relations—a view that persists to the present day in theoretical physics, where reality is held to be describable in mathematical terms and equations.

Plato appropriated this Pythagorean perspective in his ontological and epistemological hierarchy of levels of greater and lesser being that are accessed by greater or lesser modes of knowing and belief. This picture of the nature of reality and our knowledge of it was set out most famously in Plato's *Republic*, in the sections on the Allegory of the Cave (Book VII) and the Metaphor of the Divided Line (Book VI). Recall the allegory (*Republic* 514ff.), as illustrated in figure 2.1.

Men dwell for their entire existence in a dark cave, fettered so that they can only view the wall at the bottom of the cave. Behind them burns a fire,

Figure 2.1
Plato's Allegory of the Cave.

and between them and the fire lies a wall upon which people move and carry objects. The shadows cast upon the cave wall constitute the sole perceptual content of the cave dwellers. For these cave dwellers, those shadows constitute reality. Imagine, then, that one of the fettered men is released from his bonds and turns around to face the fire. He will be blinded and confused, but gradually his eyes will adjust to a reality he had not before imagined possible as he learns to see physical objects, people, and the fire. If he tries to relate this world to the captives below, they will deny it and heap scorn upon him. If the newly released cave dweller were then to be dragged up out of the cave to the sunlit world outside, he would again suffer temporary blindness and disorientation before adjusting to this new world of light. Finally, he would realize that the sun, the ultimate source of all light, is, metaphorically, the ultimate source of our knowing. Socrates summarizes for Glaucon the meaning of the allegory: "The realm of the visible should be compared to the prison dwelling, and the fire inside to the power of the sun. If you interpret the upward journey and the contemplation of things above as the upward journey of the soul to the intelligible realm, you will grasp what I surmise since you were keen to hear it" (*Republic* 517b, Grube translation, 170) (Grube & Reeve 1974).

Plato appropriates and extends the common conceptual metaphor KNOW-ING IS SEEING. Ideas are understood metaphorically as visible physical objects, knowing is metaphorically seeing the features of an idea-object, and the "light of reason" shining on idea-objects is what makes knowledge possible. The one who possesses genuine knowledge, rather than mere opinion, "sees" (i.e., via the metaphor, *knows*) the forms or essences that define objects and events. Coming to genuine knowledge is metaphorically conceived as movement from mere physical seeing of perceptual objects to the intellectual vision of the ultimate essences that make things what they are.

This account is based on an ontological dualism of the visible versus the intelligible realms. Socrates had earlier, at the end of Book VI, employed the Metaphor of the Divided Line to explain how various levels of knowledge operate. Imagine a line divided into unequal parts, and let the top part stand for the intelligible realm and the bottom for the visible. Then, divide each of the two unequal parts again, using the same proportions as the original cut. We then have four regions that decrease in size proportionally from top to bottom (see figure 2.2). The four different sizes of the levels are meant to represent the degree of reality of the objects in that domain and

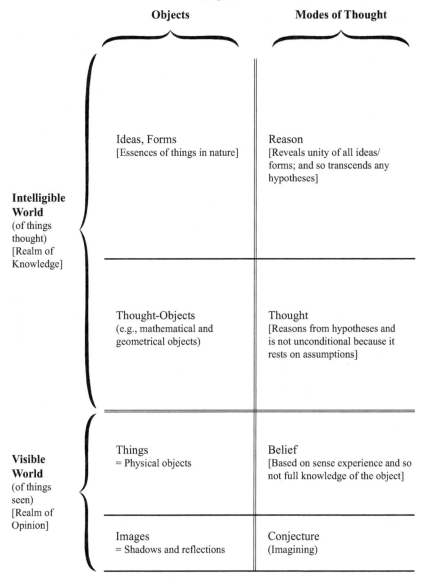

Figure 2.2
Plato's Metaphor of the Divided Line.

the corresponding degree of knowledge possible at that level. Given these four unequal domains, we can thus indicate, on the left, the four levels of mental processes (relative to knowing) and, on the right, four corresponding levels of objects known.

At the lowest level in the visible realm (which also corresponds to the lowest level of the cave), we have our capacity for image making or imagination (*eikasia*), which grasps mere images, shadows, and reflections of things. One level up in the visible is opinion and belief (*pistis*), which gives us an awareness of objects of sense perception. Because these first two levels concern changing objects of the perceptual world, they give us belief and opinion but not knowledge. Moving up to the third level, *dianoia*, we reflect on the realm of perceivable physical objects and enter the lowest level of the intelligible realm, where we use our rational capacity for mathematical and scientific thinking. This is a level of genuine knowledge, but it always remains hypothetical, perspectival, and partial, since any mathematics or science rests on some set of assumptions. At the highest level, we supposedly employ our intellect (*nous*) to grasp the forms or essences of things, not by inferential reasoning but by a sort of direct grasp (a metaphorical seeing) discerned through a process of dialectical argument. And if, per impossible for humans, we could see the ultimate and total relation of all forms and essences in one vast metaphorical vision, we would grasp what Plato calls the ultimate Form of the Good.

The key point in reminding ourselves of these two closely related Platonic models of knowledge (the Cave and the Divided Line) is to appreciate how they depend on a rigid demarcation between the visible and intelligible realms that supports a view of knowledge as pertaining only to that which is fixed, complete, and eternal—the forms or ideas (essences) of things accessed in the intelligible realm.

Plato's theory also supplied a notion of *degrees of being* forming a hierarchy from nonbeing to contingent beings and on up to necessary and unchanging being, which then supposedly corresponds to correlative levels of knowing, from mere imagining up to grasp of the unchanging forms. There thus arose the strange (to our contemporary sensibilities) notion that there are degrees of reality, such that some objects or entities possess more or greater being (or reality) than others. This view was later incorporated into Christian (esp. Thomistic) theology. God, as ultimate reality, is the creator and sustainer of all that exists. Humans, on the other hand, get both their essence (what they are) and their existence (act of existing) from God. So, they are doubly dependent on God as Being itself, or pure act of existing. In this Great Chain of Being, animals fall below humans in being non-rational and less self-actualizing, and inanimate objects fall at the bottom of the scale as passive entities lacking most of the excellences of active, self-moving, and rational beings. For humans, it is our intellect, not our body, which makes us most God-like, insofar as we are able to grasp the unchanging essences that make things what they are.

It was a short step from the Pythagorean focus on the mathematical relations of objects to the Platonic postulation of two different realms of being, "a higher realm of fixed reality of which alone true science is possible and of an inferior world of changing things with which experience and practical matters are concerned" (Dewey 1929/1984, 14). Knowledge in its full sense was of the eternal essences, while knowledge in a lesser sense (as mere opinion and belief) was of the characteristics of objects subject to change. Within this classical framework, to know something, then, is not merely to encounter the ways it appears to your sensible, physical, ever-changing body, but rather to grasp in the intellect the form or unchanging essence of what the thing really is.

As a result of this radical bifurcation, there followed the correlative split between knowing as the theoretical grasp of eternal forms and the practical belief appropriate for conducting our mundane affairs. "To these two realms belong two sorts of knowledge. One of them is alone knowledge in the full sense, science. This has a rational, necessary and unchanging form. It is certain. The other, dealing with change, is belief or opinion; empirical and particular; it is contingent, a matter of probability, not of certainty" (Dewey 1929/1984, 17).

2.3   Aristotle's Bifurcation of Theoretical versus Practical Knowledge

Aristotle appropriates the Platonic assumption that knowledge in the fullest sense must be of what is unchanging and exists of necessity.

> We all suppose that what we know is not even capable of being otherwise. . . . Therefore the object of scientific knowledge is of necessity. Therefore it is eternal; for things that are of necessity in the unqualified sense are all eternal; and things that are eternal are ungenerated and imperishable. . . . Scientific knowledge is, then, a state of capacity to demonstrate. (Aristotle 2009a)

Aristotle's world is a cosmos in which things have essences that make them the kind of thing they are and that determine their behavior. Essences are not merely convenient ideas we have of what things are. Rather, those essences are alleged to actually exist in the causal structures of the world. Therefore, in knowing the essence of a thing, we come to know what the thing really is and how it operates in the world.

Aristotle's account of theoretical knowledge rests on the assumption that our scientific knowledge of a given object is knowledge of the kinds in which it participates. We know the thing through the universals of which it partakes. So, to know a particular person in the proper scientific sense would be to know how that person incorporates and manifests all of the essential kinds of which that person is a part. For example, we know (scientifically) a certain friend as a man, son, father, husband, grandfather, professor, basketball player, fisherman, author, and so on. To know an individual is to understand the relations of the kinds that conjointly determine its nature and behavior, and those kinds are objective realities manifesting the specific essence of each kind. We can then imagine a fixed and completed universe consisting of a vast taxonomy of preexisting essential kinds, so that to know what a thing is requires understanding, first, what its nature is (what essential kinds it falls under) and, second, how it stands in relation to other kinds of things.

In several books, especially the *Prior Analytics*, the *Posterior Analytics*, and the *Metaphysics*, Aristotle argues that the demonstrative syllogism is the appropriate vehicle for acquiring scientific knowledge, insofar as it reveals the relations existing among essential kinds (Aristotle 2009b). A syllogism is a formal structure that makes it possible for us to set out the various relations that exist among kinds of things and events. The demonstrative syllogism gives us knowledge of the objective characteristics or attributes of a given entity. To establish a syllogism, we must start by identifying the appropriate premises that come to be related by various patterns of reasoning. Aristotle claims that we construct the relevant premises through observation of previous cases through intuitive reasoning, and then the syllogism reveals or demonstrates the objective relations between fundamental concepts of kinds. For example:

All As are Bs

Object O is an A

Therefore, Object O is a B

We learn what a thing is, and how it will behave, when we learn the essential kinds it manifests. Metaphorically, each essence or kind is a container in which a particular object can be placed, for example as when "John" (the individual) is "contained in" the concept *man*, so that "John is a man." Aristotle's logic is thus container logic, in accordance with the metaphor CATEGORIES ARE CONTAINERS (Lakoff & Johnson 1999). The various valid syllogistic forms allow us to explore possible relations between kinds of things. "Scientific knowledge is judgement about things that are universal and necessary, and the conclusions of demonstration, and all scientific knowledge, follow from first principles (for scientific knowledge involves apprehension of a rational ground)" (*Nicomachean Ethics* 1140b).

Aristotle's conviction in the power of demonstrative reasoning is grounded on his assumption that essences define necessary attributes of objects, and then syllogistic reasoning traces out the necessary connections among such attributes: "Demonstrative knowledge must rest on necessary basic truths; for the object of scientific knowledge cannot be other than it is. Now attributes attaching essentially to their subjects attach necessarily to them: for essential attributes are either elements in the essential nature of their subjects, or contain their subjects as elements in their own essential nature" (*Posterior Analytics* 74b). *So, for Aristotle, scientific knowledge gained through demonstrative reasoning puts us in touch with the essential nature of things and maps the rational order of nature. Any other way of relating to objects and events is an inferior mode of engagement that never rises fully to the level of fixed and eternal knowledge.*

Besides theoretical knowledge of unchanging essences, there is only the realm of practical knowing, which pertains to objects in the contingent world of perception and action. This claim is based on the distinction between the invariable and the variable in nature, each with its corresponding mode of knowing: "And let it be assumed that there are two parts [of the soul] which grasp a rational principle—one by which we contemplate the kind of things whose originative causes are invariable, and one by which we contemplate variable things. . . . Let one of these parts be called the scientific and the other the calculative" (*Nicomachean Ethics* 1139a). Since practical or calculative reasoning deals with the variable—with things that admit of possibly being other than they are—we cannot have absolute knowledge of them. Hence, concerning practical reasoning about human affairs, Aristotle cautions that we must not expect the precision we require and demand from scientific knowledge (1094b). He concludes that practical wisdom, which is the capacity to deliberate well about what is good and expedient in respect of life in general, cannot involve scientific demonstration:

> Now no one deliberates about things that are invariable, nor about things that it is impossible for him to do. Therefore, since scientific knowledge involves demonstration, but there is no demonstration of things whose first principles are variable (for all such things might actually be otherwise), and since it is impossible to deliberate about things that are of necessity, practical wisdom cannot be scientific knowledge. (*Nicomachean Ethics* 1140a)

Notice Aristotle's fateful and highly problematic conclusion: the highest form of knowledge of the world (through scientific demonstrative reasoning) is separate and different from the wisdom that helps us to lead meaningful and intelligent lives. We will see in later chapters how much this divorce of knowledge from meaning for our lives has persisted down to the present day and has reinforced inadequate accounts of how we are able to know something. We are left, then, with an ontological split between the fixed realm of eternal essences and the messy, variable world of contingent human actions and practices. Scientific knowledge is a theoretical grasp of these fixed essences and their relations, whereas practical reasoning deals with "things which are only for the most part true" (*Nicomachean Ethics* 1094b). To many, such a distinction will seem perfectly obvious, since it constitutes a recurring folk theory of being and knowing in Western intellectual traditions. However, the consequences of drawing this line between the invariable and the variable have been devastating in Western epistemology because they attribute a fixity and necessity to certain realities that subsequent empirical research will reveal to be false or at least highly questionable. As we will see, the belief that one can gain absolute knowledge of anything leads to a fundamentally mistaken conception of human knowing—a conception that assumes a radical split between theory and practice, knowing and doing. Our argument for knowing as embodied, enactive, and transformative will emerge as we develop our biologically based theory of mind.

2.4   Descartes and the Quest for Certainty

The idea that knowledge in the proper sense—knowledge of the essential being of things—concerns objects that are what they are by necessity is often closely aligned with the idea that we can have certain, indubitable, unshakeable knowledge of those things. Not surprisingly, those who claim that certain and absolute knowledge is possible have often modeled it on their conception of mathematics, which they believe to be the ultimate, universal language for the description and explanation of being. Since the seventeenth century, the most famous and influential version of this assertion has been René Descartes's (1595–1650) argument that it is possible to attain knowledge that cannot be subject to doubt. In *Rules for the Direction of the Mind* (1628) and *Discourse on Method* (1637), Descartes expresses his disappointment that all of the claims to absolute knowledge he had surveyed, over the course of his education, in all fields of human endeavor, were never able to make good on their claims to unquestionable foundations. Of philosophy, he says "seeing that it has been cultivated for many centuries by the best minds that have ever lived, and that nevertheless no single thing is to be found in it which is not subject of dispute, and in consequence which is not dubious, I had not enough presumption to hope to fare better there than other men had done" (*Discourse on Method*, 1637/1970, 85–86).

Undaunted by this long history of failed attempts to find a method capable of guaranteeing certain knowledge, Descartes turns to his training in mathematics for the clue to a new starting point: "of all those who have hitherto sought for the truth in the Sciences, it has been the mathematicians alone who have been able to succeed in making any demonstrations, that is to say producing reasons which are evident and certain" (*Discourse on Method*, 1637/1970, 92–93). Descartes's method is to subject all claims to knowledge to critical examination, in search of something which is beyond all doubt and can be regarded as necessary and certain truth. His procedure was "to accept nothing as true which I did not clearly recognize to be so: that is to say, carefully to avoid precipitation and prejudice in judgments, and to accept in them nothing more than what was presented to my mind so clearly and distinctly that I could have no occasion to doubt it" (92). Psychologically, this amounts to transcending any anxiety that might arise from indeterminacy and ambiguity in experience. This is an impossible task because, as we will see, the anxiety of doubt is built into our deepest systems of motivational control.

The starting point for all knowledge is what Descartes calls the mental operation of *intuition*, which is an unmediated grasp of the truth of some idea or proposition.

> By intuition I understand, not the misleading judgment that proceeds from the blundering constructions of imagination, but the conception which an unclouded and attentive mind gives us so readily and distinctly that we are wholly freed from doubt about that which we understand. Or, what comes to the same thing,

intuition is the undoubting conception of an unclouded and attentive mind, and springs from the light of reason alone. (*Rules for the Direction of the Mind*, 1628/1970, 7)

Notice that although Descartes took this account of intuition to be a literal truth, his entire perspective depends on a common conceptual metaphor of KNOWING IS SEEING, in which we conceive of acts of understanding and knowing metaphorically as acts of visual perception based on the following mappings across the source domain of vision and the target domain of knowledge:

The KNOWING IS SEEING Metaphor

- Ideas Are Objects
- Knowing Is Seeing
- Reason Is A Natural Light
- Intellectual Acuity Is Visual Acuity
- Intellectual Confusion Is Blockage or Impediment to Seeing Something

According to the KNOWING IS SEEING metaphor, if some idea-object is viewed by the mind's eye in sufficient light of reason, then we cannot help but know that idea. The metaphor-based logic is precise, and Descartes follows it out in every detail, concluding that "our inquiries should be directed, not to what others have thought, nor to what we ourselves conjecture, but to what we can clearly and perspicuously behold and with certainty deduce; for knowledge is not won in any other way" (*Rules for the Direction of the Mind*, 1628/1970, 5).

Intuition alone is not sufficient for building up a foundation of certain knowledge. As an act of quasi-vision, it gives us some true and certain propositions, but we have to connect those propositions into chains of reasoning that are themselves truth preserving and immune to doubt. Descartes calls this reasoning process *deduction*, by which he means "all necessary inference from other facts that are known with certainty" and which "are deduced from true and known principles by the continuous and uninterrupted action of a mind that has a clear vision of each step in the process" (*Rules for the Direction of the Mind*, 1628/1970, 8). Notice that deduction introduces a temporal dimension into reasoning that is not present in the intuitive "seeing" of a proposition as true in a single act (of metaphorical vision) at a point in time. Descartes understands deduction as a stepwise process of moving from one intuitively envisioned idea to another, and this introduces a different metaphor, namely, DEDUCTION IS STEPWISE MOTION ALONG A PATH (see Lakoff & Johnson 1999, ch. 12, for a fuller analysis of Descartes's metaphorical conception of knowledge). Consequently, Descartes ends up claiming that reliable deduction consists of running over in the mind inferential steps (each of which is intuitively clear) so quickly that they asymptotically approach something like a single momentary intuitive vision of all the connections. The clash of these two foundational metaphoric assumptions (i.e., INTUITING IS SEEING versus DEDUCTION IS STEPWISE MOTION ALONG A PATH) is based on the ontological and epistemological discrepancy between an instantaneous act of intuition versus a temporally extended operation of deductive inference. Descartes, recognizing this problem, tries to turn deduction into an instantaneous vision of intuitive connections:

> For this deduction frequently involves such a long series of transitions from ground to consequent that when we come to the conclusion, we have difficulty in recalling the whole of the route by which we arrived at it. This is why I say that there must be a continuous movement of thought to make good this weakness of the memory. . . . To remedy this I would run over them [the inferential connections] from time to time, keeping the imagination moving continuously in such a way that while it is intuitively perceiving each fact it simultaneously passes on to the next; and this I would do until I had learned to pass from the first to the last so quickly, that no stage in the process was left to the care of the memory, but I seemed to have the whole intuition before me at one time. (*Rules for the Direction of the Mind*, 1628/1970, 19)

We present Descartes's metaphor-based account of foundational knowledge as "seeing" some eternal object as representative of the recurring desire to find something so necessary, so certain, and so unchanging that, once clearly grasped, it can never be doubted. There is no such thing, but that has not stopped people from yearning for something beyond all time and place, (established by God or Nature or Reason) that manifests an ultimate rational order underlying all reality.

It is a significant irony that Descartes, who thought he was offering a literal, disembodied account of mind and knowledge, in fact has a theory that is entirely dependent on metaphors that understand intellectual cognition in terms of bodily perceptual and motor activities. In chapter 6 we will argue that such body-based metaphors are the stuff of abstract understanding and reasoning generally, so that any account of mind and knowing must inevitably employ *some* particular metaphor that is grounded in bodily experience. Even our most extreme claims about disembodied, objective truth are meaningful only because they are rooted in bodily experiences and processes.

It is no accident that Descartes turned to mathematics and logic, as so many before and after him have done, to look for absolute foundations for knowledge. In the early twentieth century, it was the mathematician and logician Gottlob Frege who proposed something like a Platonic realm of objects that were supposed to provide the basis for the possibility of objective meaning and truth. Assuming that neither bodily processes nor mental operations could give the appropriate universal basis for objective knowledge, Frege postulated a third realm of being (to which he gave no name), consisting of abstract entities such as numbers, concepts, propositions, mathematical and logical functions, and what he called The True and The False. Although we cannot discuss Frege's objectivist theory of mind and language here (see Lakoff & Johnson 1999, ch. 21 for a fuller analysis), it is worth noting that Frege is typically considered to be one of the founders of analytic philosophy, and his Platonism has therefore often shaped subsequent theories of mind, knowledge, and language to the present day.

In *Where Mathematics Comes From: How the Embodied Mind Brings Mathematics into Being*, George Lakoff and Rafael Núñez describe and criticize what they call "The Romance of Mathematics" (2000, 339. This is roughly the view that mathematics (and logic) constitute a universal language of thought that defines the essence of human rationality, that reality (Nature) manifests this rational order, and that mathematics gives us ultimate access to the very nature of things (339ff.). According to this view, mathematics is not just in the mind. Rather, it is realized in the structures, relations, and processes of nature. Lakoff and Núñez carefully and thoroughly dismantle this divinization of mathematics, but they are under no illusion that those enamored of such a view will ever cease to see mathematics and logic as the only subject where certainty and universality are possible. *Where Mathematics Comes From* is an account of how various mathematical systems, like *all* human conceptual systems, are constructed out of body-based meaning structures such as images, schemas, and conceptual metaphors. As beautiful and important as mathematics is, it cannot lay claim to disembodied truth or modes of reasoning. The greatest achievements of mathematics in understanding our world are precisely the result of their rootedness in and utilization of sensory and motor processes and structures.

2.5   The Kantian Turn: Philosophy as Epistemology

In some important respects, Immanuel Kant's account of knowledge represents a welcome turn away from the Cartesian quest for rational certainty. However, Kant still leaves us with a view that fails to appreciate the embodied, affective, and value-laden character of human knowing. He also encourages the view that philosophy ought to develop an adequate theory of cognition and knowledge without room for the subjective dimensions of meaning and value. This epistemological focus defines Kant's Critical Philosophy, which investigates the nature, conditions of possibility, and limitations of various human modes of thought and judgment. Each of Kant's three great critiques takes up a particular type of judgment that lays claim to universal validity and then attempts to explain how that type of judgment is possible, given what he claims about how the mind works. The *Critique of Pure Reason* (1781/1968) tries to show how empirically grounded universally valid knowledge judgments of the sort relevant to a scientific understanding of our world are possible. The *Critique of Practical Reason* (1788/2002) examines how moral judgments and laws claiming universality are possible. The *Critique of Judgment* (1790/1987) examines how judgments of natural beauty, sublimity, and teleology are possible and can claim universality, even though they appear to be based on feelings, rather than on shareable concepts.

Kant is typically praised for having dramatically advanced our understanding of how knowledge is possible because he gives up the idea that we can know reality as it is in itself, and recognizes that we can only know things as they appear to us through the mediation of our perceptual organs and the organizing conceptual patterns and processes of our minds. In the *Critique of Pure Reason*, he develops his seminal idea that we can only know a thing as it is given to us in experience (through perception) and thought by means of concepts supplied by the universal formal structures of our minds. We supposedly can have genuine shared objective knowledge of nature because all humans have similar sensory capacities and structure their thought via concepts that can be shared and communicated to other creatures like us, who have the same general physical and mental makeup.

This new conception of the possibility of knowledge, or "empirical cognition" (*empirische Erkenntnis*), does not traffic in claims about certainty, but it makes extensive use of the notions of universality, necessity, and purity of certain types of concepts and forms of judgment. Kant's sensibilities seem more modern and contemporary because he claims that we can know our world precisely because we have had a hand in constructing it, through the filtering mechanisms of our perceptual faculties and the form-giving and organizing activities of human understanding and reason.

Aristotle, as we saw, thought that objective scientific knowledge was possible primarily because the forms of things known through thought exist, objectively, in the world. They are not mere operations of mind. Thus, Aristotle thought of categories (i.e., the ultimate characteristics of any object in nature) as existing in the world, but also as being graspable by human intellect. Kant, in contrast, saw categories (i.e., the basic characteristics of any object of possible experience) as resulting from "pure" concepts of understanding possessed by all humans and therefore constituting the conditions for our ability to know an object. Kant's revolutionary claim was that we can know this category structure of our world *because we constructed it* via innate concepts shared by all adult humans. Kant's categories—such as cause and effect, unity, reality, and intensive magnitude—result from the types of logical judgments we humans make, whenever such judgment types are applied to any object of possible experience. In short, our world manifests knowable category relations because such relations are imposed by human understanding as conditions for experience of any object.

What's not to like in this human-centered account of knowledge? Kant was right to recognize that knowing is always mediated through the structures of human understanding. However, Kant thought some of those mediating structures had to be disembodied—independent of the particularities of human brains and bodies. Kant insists that what makes something knowable are the concepts (both empirical ones arising from experience and pure ones imposed by the structures of rational understanding) under which it falls. Concepts provide the knowable form of any particular thing, and they result from mental operations of understanding and reasoning. Kant argues that humans can access "pure" concepts of understanding untainted by anything bodily and thus not derived from experience. These pure concepts are based only on the form-giving capacities of our mind, grounded ultimately, he claims, in an organizing activity of a transcendent ego that is capable of unifying and ordering activities of thought. Understanding combines concepts into propositions, and then reason discerns inferential relations among propositional judgments to build up larger meaningful wholes (gestalts) that constitute knowledge. Despite his insistence on experiential input for any process of empirical knowing, Kant never abandons an intellectualizing approach to knowing that assumes pure (nonempirical) concepts and reasoning. In the final analysis, for Kant, it is never emotions, feelings, or values that determine the content and character of our knowing activities Rather, it is the form-giving, universal structuring activity of the pure ego and the empirical ego that underlies all knowledge.

Over the last two centuries, Kant's account of knowledge has been so influential that, especially throughout the twentieth century, it defines the core of philosophy—so much so that Richard Rorty, in *Philosophy and the Mirror of Nature* (1979), credits Kant with turning philosophy into epistemology. Modern philosophy, at least since Descartes, became increasingly concerned with the conditions for the possibility of knowledge, especially knowledge that is objective and universally valid. One way to read the history of modern philosophy is as a series of attempts to define methods of inquiry to secure objective knowledge of nature, including human nature. Descartes claimed to find certainty in the inner world of consciousness. As Rorty observes, Kant eventually came to treat the outer world of physical nature (including our human bodies) as the product of our human structures of sensory perception and conceptual ordering, rather than things as they are "in themselves." We can know the world because of our role in filtering perceptual inputs and cognizing them through shared concepts imposed by mind. The self that does this cognitive ordering and reasoning is taken by Kant to be nonempirical, lying beyond and behind our bodily experience in the phenomenal world.

Rorty sums up this fateful philosophical development as follows:

> Kant put philosophy "on the secure path of a science" by putting outer space inside inner space (the space of the constituting activity of the transcendental ego) and then claiming Cartesian certainty about the inner for the laws of what had previously been thought to be outer. He thus reconciled the Cartesian claim that we can have certainty only about our ideas with the fact that we already had certainty—a priori knowledge—about what seemed not to be ideas. (Rorty 1979, 137)

In other words, Kant denied knowledge of things in themselves, but he secured the possibility of shared knowledge by regarding the outer world perceived by our senses as knowable just insofar as it was structured by concepts arising from the formal character of the inner world of thought. Kant understood that through interaction with what is "not us," we co-constitute our experience and give it its character as known, but he took the self (as transcendent ego) that does this organizing to be unknowable because it is not bodily. Ultimately, he is forced to treat the nonempirical self as a nonexperiencable source of spontaneous organizing activity. We will later examine extensive neuroscientific evidence explaining the possibility of reliable human knowledge emerging from the biological domain, but not requiring any such transcendent ego working behind and beneath consciousness to unify our experience.

2.6   The Linguistic Turn Away from the Body

Rorty is also well known for having appropriated Gustave Bergmann's description of twentieth-century Anglo-American analytic philosophy as having taken a decisive "linguistic turn" (Rorty 1967). The central idea was that all meaning, thought, reasoning, and knowledge are linguistic in character and tied to language practices. So, it is through language (its syntactic features, semantic contents, and pragmatic uses and effects) that we access thought and everything connected with it. Although language has clearly transformed the capacity for human thought, analytic philosophers, like early cognitive scientists, conflated the mind with its linguistic form.

Instead of focusing on the structures and processes of experience as manifestations of the full domain of biological intelligence, analytic philosophers turned everything into an analysis of linguistic concepts and propositions and the formal procedures for ordering and relating them. Frege had fatefully shaped subsequent philosophy of mind, language, and knowledge by insisting that the fundamental unit of thought was the proposition, with its subject-predicate structure that mirrors that same structure found in our linguistic sentential speaking and writing. The proposition, Frege claimed, is the unit that maps (or fails to map) onto some state of affairs in the world, thereby constituting a truth claim. We understand the sentences we hear because we can grasp the underlying concepts and propositions referred to by those sentences. Linguistic analysis supposedly gets us to the underlying processes of meaning, thought, and knowing. Knowledge is modeled on linguistic structures accessed by intuitive judgments of language users. The body gets almost entirely left out in favor of armchair analysis based on linguistic intuitions about syntax, semantics, and pragmatics.

We have come full circle to where we began this chapter, with the long quotation from the *Stanford Encyclopedia of Philosophy*. This is philosophy as epistemology and a very limited view of epistemology at that. It ignores most of the processes and functional organizations of embodied creatures that operate in human modes of knowing. In the effort to reduce the process of knowing to explicit rational propositions, analytic philosophers confuse knowing the world with their linguistic constructions. As John Dewey suggested, they commit *the philosopher's fallacy* of thinking that their rational, linguistic descriptions of things are the things themselves, or at least that their preferred descriptions capture everything that matters, with respect to knowledge. What is missing from "S knows that p" epistemology is any deep and serious exploration of the role of our embodiment, our emotions, our motivational processes, and our values. In other words, what is missing is an adequate account of the role of the embodied self in knowing, to which we now turn.

3 The Intertwining of Self and Knowledge

What and how we know is partly, though significantly, constitutive of who we are. We say "partly constitutive" because the self is much more than a knowing subject; it revels in the qualities of things, relishes food, seeks activity, labors to make a living, enjoys natural beauty, argues, makes love, dances, loses itself in music, and shares life with others, to name just a few of its defining characteristics. That said, the activity of knowing is a crucial part of our ever-changing self-identity. Cognition is learning, and this process extends over a person's life span. Insofar as learning changes our brains and bodies, it results in a transformed self. Knowing is a process that arises from our self as it is presently constituted, and it then eventuates in a transformation of our identity into a new self. Since knowing is an ongoing, developmental affair, our selfhood is continually being remade over the course of our lives through active engagement with the world.

The idea that self and world are interdefined, co-constituted, and aspects of one and the same developmental process is not our commonsense view, nor has it been the dominant view in Western philosophy. For the last 2,500 years, knowing has mostly been viewed as adding new conceptual or propositional content to a preformed and fixed mind or self—a self that then judges or evaluates how that content relates to the world. There has been a tendency to think of the self as a fixed or highly stable structure that exists prior to its acts and that then takes in information, evaluates it, and initiates action. Mind is regarded as a central processing unit that receives representations from the outside world, conceptualizes that sensory input, and rationally assesses it. On this view, the self is not changed, in its essential nature, by what it knows. Rather, it is both the source of our receptivity of sensations and the active agency that orders and evaluates what has been perceived.

This conception of mind posits inner representations that must some-how mirror mind-independent realities if they are to constitute knowledge. Knowledge then becomes primarily a matter of how the inner process con-nects with, mirrors, or re-presents something that is physical and "outer"—out there in the mind-independent world. Knowledge becomes a matter of reporting how things are as opposed to a process of changing how things are. Based on the alleged separation of mind and body, mind and world, the problem of knowledge reduces to determining how we can know whether what a proposition expresses is true or false.

Through the end of the nineteenth century, this conception of mind at least gave the self a significant role in the knowing process. However, in the twentieth century, in Anglo-American analytic philosophy, even the self was mostly eliminated from the knowing relation. Theory of knowledge focused on how a proposition could refer to a mind-independent state of affairs. There was a proposition, a mind-independent world, and an objective rela-tion between the two. Notice that there is no mention of the "self" in this relation. Truth was taken to be an objective relation, not dependent on the nature of the knowing self. Philosophy of mind and language focused almost exclusively on linguistic structures as the locus of meaning and thought (Frege 1966; Quine 1960; Rorty 1989, 3–22). Consequently, theories of knowl-edge and truth were about the conditions under which a proposition could be evaluated in terms of its relation to objective states of affairs in the world. As we will see in chapter 5, the behaviorist psychology that dominated the first half of the twentieth century eschewed reference to mind, inner states, and feelings, focusing mostly on stimuli and consequent behavior, supposedly without need of subjective processes. This reinforced the philosophical epis-temology that tried to minimize, or even deny, any role for inner processes underlying cognition.

What these theories were *not* about was how knowledge could be subjec-tively meaningful for our lives. In fact, the whole idea was to keep unwanted subjectivity out of the equation because it was notoriously difficult to study and impossible to quantify. Ideally, the meaning of a proposition, and its knowledge status, was an objective relation between the proposition and some state of affairs in the world. Within this framework, knowledge could supposedly be explained without any mention of subjective processes or realities. Mainstream analytic philosophy mostly tried to keep subjective factors—meaningfulness, emotions, values, and motivations—out of the discussion of meaning, thought, and knowledge.

We will provide evidence that the subjective dimensions are crucial to human knowing. Knowing is not only about objectively representing our world; it is equally about the meaning for our lives of what we experience.

These dominant theories of knowledge assume classic ontological and epistemological dualisms, such as mind and body, subject and object, inner and outer, and thought and feeling. As the pragmatist philosophers argued, if you regard these opposing terms as representing radically separate and distinct realities, then you will never be able to get them back together again. That is, if, on the one hand, you think of knowing as something that goes on only within an inner mental realm, you will never be able to explain how what is inner (a representation) can relate to what is outer (a mind-independent situation). If, on the other hand, you regard knowledge and truth as relations of propositions to objective states of affairs, leaving out the person who knows, you will never be able to explain how that knowledge is meaningful, nor will you be able to explain how a knower transforms its world, even as the knower is being transformed by it.

The pragmatist solution to this problem, as we will see in the next chapter, is to realize that the subject as knower and the object known are not two separate or ontologically different kinds of reality that have to become miraculously joined together. Rather, they are co-constituted aspects of one and the same experiential process of organism-environment transactions.

Our claim is that knowing is not primarily a mirroring or representational act, but rather a constructive activity that transforms both our selfhood and our world. It is more a doing and making of new states of affairs than it is a reporting on preexisting states of affairs. According to the biological pragmatist perspective that we develop here, if you start with one continuous process of organism-environment interaction, then the knower and the known emerge as co-constituting aspects or dimensions of that experiential process. Knowing gets reconceived, not as the mind mirroring what is not mind, but rather as an activity by means of which the knower is reconstructed and simultaneously remakes their world. These transformations expand and enrich the meaning of experience, thereby often improving the knower's ability to function in his/her surroundings. If you don't make the mistake of extracting the self as knowing subject from the world as known object, then you don't face an ontological/epistemological gap that later needs to be bridged and that requires problematic notions such as mirroring, correspondence, and referring.

3.1 The Relation of Self and Knowledge in Modern Philosophy

For the last twenty-five centuries, philosophers have looked for some quality of an idea or mental activity that could guarantee that it is a faithful representation of some piece of the world. They have sought to characterize the mirroring, correspondence, referring, representing, or mapping relation obtained between ideas in the mind and things in the world when we "know" something. That is the formulation of the knowledge problem one gets by radically differentiating the knower and the known.

To see what's wrong with this picture, and to appreciate the pseudo-problems that it generates, we consider two famous philosophical accounts of knowing. One is René Descartes's rationalist metaphysics of mind that posits a knowing self that precedes, and is left unchanged by, what it knows. The other is David Hume's empiricist theory, which draws skeptical conclusions from its failure to find just such a substantial knowing self. That is, knowledge requires a prior self that knows. So, if there is no evidence for a substantial self, then the possibility of knowledge is called into question, and we are left with skepticism.

We suggest that although these two celebrated views might appear to be radically opposed, they share the assumption that the possibility of knowledge requires a fixed self that entertains representations of external realities. Assuming such a self, Descartes leads us on a quest for certainty, while Hume finds no such self, and is thereby thrust into a skepticism that denies the possibility of certain knowledge. We examine Kant's self-described role as unifier of these two perspectives (i.e., rationalist and empiricist strains) by recognizing that we can know the world because we have made it in our own image. Kant saw that subject and object are co-constituted, so they are not independent realities. The unity of the self is realized as the unity of objects in experience. Kant's account is inadequate to the extent that it posits a transcendent ego that makes knowing possible, but he correctly saw that self and object are correlative realities. Appreciating how knowledge and identity are intertwined, we will later examine, from a neuropsychological perspective, the inherent transformations of the self that are implied by the neural organization of meaningful concepts.

3.2 Descartes's Fixed, Pre-given, Disembodied Self

From Descartes (1596–1650) to the end of the nineteenth century, modern philosophy obsessed over the conditions that make objective knowledge possible. Descartes started by insisting that knowledge had to be certain and indubitable. *"Only those objects should engage our attention, to the sure and indubitable knowledge of which our mental powers seem to be adequate"* (*Rules for the Direction of the Mind*, Rule II, 3). This stringent demand for certainty leads him to "reject all such probable knowledge and make it a rule to trust only what is completely known and incapable of being doubted" (3).

Descartes took the mark of indubitable, certain knowledge to be the clarity and distinctness of an idea or thought. Recall that in Descartes's system, intuition—the instantaneous (metaphorical) "seeing" of an idea with clarity and distinctness—needs to be combined with deduction, by which we create new knowledge by inference from already-known propositions. Whereas intuition is an instantaneous quasi-visual act of cognition, deduction is a process of connecting thoughts over time, and has an intrinsic temporal character that is seemingly absent in immediate intuition. Consequently, we saw how Descartes was led to propose that in order to capture the temporal and serial order of deduction, we should practice running through a sequence of thoughts (i.e., observing in an intuitive glimpse the connection of thought A to thought B, and then intuiting the connection of thought B to thought C, and so on) over and over again so quickly that what are actual distinct temporally different individual acts of intuition merge into a single instantaneous intuitive act.

Descartes's founding KNOWING IS SEEING metaphor leads to a search for certainty in an act of quasi-vision and to the attempt, per impossible, to turn the temporal activity of deducing one thing from another into a single, momentary act of intuition (as clear and distinct perception). Descartes's theory of mind and knowing reveals what is wrong with trying to define knowledge in terms of self-validating inner representations. There is no property of a mental representation that could guarantee its truth.

The chief problem with Descartes's appropriation of the KNOWING IS SEEING metaphor is his interpretation of the metaphorical seer as a pre-given rational ego in full possession of its powers of understanding and reasoning (i.e., as a complete self that exists prior to its acts of perceiving, thinking, moving, and manipulating objects). His rationalistic, intellectualist conception of knowledge is based on an individual who is capable of turning inward to inspect the contents of its own mind and to perform mental operations (i.e., judgments) on those contents. From this perspective, the self possesses reflective capacity that grasps the character of an idea or thought with utmost clarity and without need of a body.

> Just because I know certainly that I exist, and that meanwhile I do not remark that any other thing necessarily pertains to my nature or essence, excepting that I am a thinking thing, I rightly conclude that my essence consists solely in the fact that I am a thinking thing [or a substance whose whole essence or nature is to think]. And although possibly . . . I possess a body with which I am very intimately conjoined, yet because, on the one side, I have a clear and distinct idea of myself inasmuch as I am only a thinking and unextended thing, . . . it is certain that this I [that is to say, my soul by which I am what I am], is entirely and absolutely distinct from my body, and can exist without it. (Meditation VI, 190)

Such a self need not rely on others to achieve knowledge, since the source of its certain, clear, and distinct ideas is its faculty of reason operating up to its full potential for reflection and thought. That rational self is not altered by its acts of knowing, but only adds, in a successive fashion, to its storehouse of known representations and their logical relations. The rational self is not changed by either its acts or the contents of its knowing, for it is our very capacity to intuit and deduce.

Descartes gave us a fixed self as a rational ego possessed of certain innate powers of thought that are not substantially dependent on bodily processes. The self knows only its inner representations, and it then has to seek some characteristic of those representations to guarantee that they map the external world (or not). The doubtfulness of our ordinary understanding leads to a measure of skepticism about the possibility of knowledge for a philosophically untutored mind. Moreover, there is no place for emotions, feelings, and values in this intellectualist account of knowledge. The errors of Cartesian claims will become more evident as we develop our alternative, embodied conception of motivated, value-based, emotionally charged, and subjectively meaningful knowing.

3.3   Hume's No Self

In the attempt to account for empirically grounded "objective" knowledge, empiricist theories fare better than rationalist theories, since they start with sensations caused by the actions of external objects on our senses, and begin with contact—or sensory interface—with the external world that Cartesianism couldn't supply. However, they brought their own problems concerning how the received sensations come to be regarded as foundations of knowledge. David Hume (1711–1776) argued that any cognitively meaningful idea must be grounded on some set of elementary sensations, which he called impressions. In *A Treatise of Human Nature* (1739), Hume divides all mental perceptions into either impressions or ideas:

> The difference betwixt these consists in the degrees of force and liveliness with which they strike upon the mind and make their way into our thought or consciousness. Those perceptions, which enter with most force and violence, we may name impressions; and under this name I comprehend all our sensations, passions and emotions, as they make their first appearance in the soul. By ideas I mean the faint images of these in thinking and reasoning. (Hume 1739, Bk. I, Pt. I, Sect. I, 1)

Notice the MIND AS CONTAINER metaphor operating here, in which the mind or soul is a container for impressions and ideas. The mind or soul is different from the impressions and ideas contained in it, and these ideas are different from the things in the world that gave rise to them. This presupposes a preexisting mind that has these perceptual experiences and that registers those experiences as contents of that mental container.

In a way that presages ideas now current in cognitive neuroscience, Hume claims that ideas are something like faint copies or images of the sense impressions and passions that press themselves upon us in perception. They are not airy abstract mental entities, but rather traces of sensory-motor-affective experiences. In the framework of contemporary cognitive neuroscience, these concepts are realized in activated neuronal clusters that are involved in experiencing the objects and events that the concepts are "of." According to Hume, the way we keep our ideas experientially grounded and constitutive of knowledge is to trace them back to their origins in simple ideas, which are, in turn, copies of simple impressions. "Simple perceptions or impressions and ideas are such as admit of no distinction nor separation" (Hume 1739, Bk. I, Pt. I, Sect. I, 3). Therefore, any complex idea needs to be broken down into its component parts—simple ideas—which can then be traced back to the corresponding simple impressions on which they rest. Voilà—knowledge of the world!

Using this method for grounding complex ideas in complexes of simple impressions, Hume asks whether we can find the simple impressions upon which two of our most important concepts—that of cause/effect relations and that of the self—could be rooted in experience. To his consternation, he claims to find no simple impression corresponding to our idea of necessary connection between cause and effect (Hume 1739, Bk. I, Pt. I, Sect. II). Even more disturbing, he can find no simple impressions corresponding to our idea of a substantial self capable of doing the knowing. He points out that when he turns inward in reflection to find his "true self," he encounters only a flow of discrete sense impressions habitually associated, instead of finding an impression of a core self or ego. When he looks for a simple impression of self, all he finds is a succession of associated impressions of body states. There is no simple impression of a unified and unifying self. "For my part, when I enter most intimately into what I call myself, I always stumble on some particular perception or other, of heat or cold, light or shade, love or hatred, pain or pleasure. I never can catch myself at any time without a perception, and never can observe any thing but the perception" (Bk. I, Pt. IV, Sect. VI, 252), Hume explains. He concludes, "I may venture to affirm of the rest of mankind, that they are nothing but a bundle or collection of different perceptions, which succeed each other with an inconceivable rapidity, and are in a perpetual flux and movement" (252).

Hume's deflationary view of the self is compatible with contemporary neuroscience evidence that there is in the brain no central executor or unifying ego. A complex set of activated neuronal clusters massively coordinates over time to orchestrate the play of consciousness, although not under the guidance of any inner homunculus who might compose the music of mind. Lacking access to the neural basis of selfhood, Hume is left with a disturbing skepticism—he does not deny that there might be a self, but he agnostically claims that we have no knowledge of such.

Descartes thinks there is a substantial self that is the seat of knowing activity. Hume thinks we need a substantial self that knows, and when he cannot find any such self, he is led to skepticism about knowledge of any substantial unified self. Hume's account of experience mistakenly posits something like raw sense impressions that arise from our contact with aspects of our environment. The problem is that he takes them to be discrete sense impressions that need somehow to be connected into ideas and thoughts, but there appears to be no self to do the collecting, synthesizing, and ordering functions. He asserts that we associate one idea with another on the basis of developed habits tied to the resemblance and/or contiguity of ideas, although he cannot find the "I" who performs this ordering activity. We are left with perceptions allegedly tied to nature, but without any self to unify, analyze, or know them.

Contemporary cognitive science research suggests that Descartes and Hume got it wrong, although Hume comes closer to an adequate account than does Descartes. Contra Descartes, there is no scientific support for positing a mental substance that does the knowing, nor is there support for conceiving of thinking as emotion and value free (Damasio 1994). Contra Hume, although the psychological evidence shows that self-awareness of our mental processes is limited, this does not undermine the possibility of genuine knowledge of our world, nor does it necessarily deny a self. If we take the rationalistic turn, we end up with a pre-given self that is independent of what it knows, though how mind connects up with nature is a mystery (to be solved, for Descartes, only by Divine intervention). If we take the Humean empiricist turn, we don't ever find a self who can know or be known. Hume's naturalistic, empiricist account of perception and thought can probably be made serviceable with the aid of today's neuroscience of perception, and his failure to observe a core self is compatible with the fact that in neuroscience we generally find no basis for, or need of, a rational ego or executive center. It is all done with distributed parallel processing. However, we need not follow Hume into his skeptical gloom because we can provide a neurally based biological and social theory of mind and knowing that requires no such rational ego. It appears that the very notion of a rational ego may be the result of assuming that mind has the (European) linguistic structure of agent-action-object.

3.4 Kant's Co-constitution of Self and World

In his *Critique of Pure Reason* (1781/1968), Kant observed that these two competing conceptions of knowledge each suffered from an overly narrow focus on one of two important dimensions of knowing to the exclusion of the other. The rationalists focused mostly on knowledge as coming from within the mind (innate ideas, pure reason, or a priori structures of cognition), while the empiricists attended only to atomistic sensations, generated as external objects affected our sensory organs. Kant's self-described achievement was to recognize that empirical knowledge requires both pre-given formal ordering structures of mind (the rationalist view) and sense perceptions created by objects acting upon us (the empiricist view). Empirical knowledge is the result of sense impressions arising as objects affect our sense organs, which are then synthesized into unified, meaningful wholes by innate structures of thought. Kant's insight was that we can know objects only as we experience them through the filters of our sensory and conceptual structures, but never as those objects exist in themselves, apart from our engagement with them. We know objects because we contributed structure and order to them. In short, we can know nature because we have partly made nature, through the ordering processes of our minds, but always subject to the affordances of our natural world.

Here, in an abbreviated form, is how Kant thought that process worked. In the *Critique of Pure Reason* he proposed that knowledge of objects requires that (1) some perceptual content be generated by our sensory organs engaging their environment; (2) this manifold of sensations has to be synthesized into a unified image (of an object or state of affairs); and (3) the object has to be thought by applying some appropriate concept to the unified sensory manifold. Knowledge, according to the "A" Deduction section of the *Critique of Pure Reason*, is the result of a threefold synthesis: the synthesis of apprehension in intuition, by which we experience a manifold of sensations as unified into an image in time; the synthesis of reproduction in imagination, by which we bind together prior unified representations with present ones, to form a unified image of an object persisting over time; and the synthesis of recognition in a concept, by which we think that object as a certain kind of thing by applying a concept to it (Kant 1781/1968, A100ff.).

It is worth looking a bit more deeply into Kant's view to understand his profound impact on subsequent theories of knowledge in order to show which parts of his view appear to be supported by contemporary cognitive science, and to explain where he went wrong on certain key issues. Kant assumed—mistakenly—that what is given by mind-independent natural sources (objects) operating on a knower's body was a flow of atomistic, unconnected sense impressions. But since we perceive unified stable objects, he hypothesized that their unity and constancy must come from the synthesizing activity of our minds, and that in two ways. First, objects have to be given to us in sensory intuition (i.e., perception). Every empirical object we are able to experience is subject to what Kant called the forms of sensible intuition (i.e., outer sense [space] and inner sense [time]). In other words, any object of experience and thought must exist *in time*, and any physical object must also exist *in space*. Space and time are not absolute metaphysical expanses, as Newton had believed, but rather are structures imposed by our bodily and mental makeup to shape what and how we experience things. Second, once we have a unified representation of an object given in perception, we only know what that object is when we are able to apply concepts to it, so as to recognize the kind of thing it is. Consequently, our experience and knowledge of any physical object is filtered through both our forms of sensibility (space and time) and our conceptual systems (both empirical concepts and pure concepts, or categories).

One additional complexity, concerning the nature of our conceptual filtering, needs to be introduced, since it constitutes Kant's rejoinder to Hume's skepticism about knowledge. Kant assumed the (erroneous) faculty psychology of his day, which claimed to explain all our cognitive operations as resulting from the interaction of discrete, independent mental faculties, such as sensation, imagination, feeling, understanding, and reason. According to this framework, the faculty of sensation is our capacity to receive atomistic sense impressions through our interaction with our environment. The faculty of imagination then organizes these disconnected sense impressions into unified images or gestalts persisting over time. Finally, understanding—the faculty of concepts—analyzes the structure of this unified perception and re-cognizes it via concepts, some of which are innate for all humans and others that are learned through sense experiences. For example, an objective experience (and its attendant knowledge) of a dog would require various sensations impressed upon us from visual, tactile, aural, and olfactory perception of Fido, unified into a figure of a four-footed, furry animal, with a wagging tail, and then conceptualized as a dog insofar as the object has all of the requisite properties constituting our concept *dog*.

Kant calls concepts learned in this manner from experience (e.g., dog, cup, tree, jump, run, fight, throw) *empirical* concepts. This is Kant's empiricist side, but he also has a rationalist side found in his notion of "pure concepts of the understanding." Pure means not drawn from experience, but rather imposed on experience as preestablished, universal formal structures that shape everything that can be an object of thought. Pure concepts are supposedly brought to experience, not derived from it, and in that sense, they are innate. Kant argues that all normal adult humans impose the same innate logical forms of judgment on the things they understand and reason about. He provides a "table of judgments" that lists twelve forms of logical relation (e.g., "if p then q," "p or q," "it is possible that," "it is necessary that") that define the essential formal patterns of thought.

These forms of logical judgment, when applied to objects of experience, become the categories, and they constitute the ultimate structures of anything we can experience. So, correlative to the table of logical judgments, Kant constructs a parallel "table of categories" that lists the twelve pure concepts applicable to anything we can call an object of experience. For example, the "if p then q" logical relation becomes, when applied to the *phenomena* of our experience, the "cause/effect" relation (as in, "If event p occurs, then event q will occur").

To counteract Hume's skepticism about the basis for cause/effect relations in the world, Kant claims that we experience cause/effect relations precisely because our understanding *imposes* that relation (as a pure concept) upon every object of possible experience. If something is a physical object, then it is necessarily embedded in causal events. Cause/effect relations are thus constitutive of *phenomena* (things as we experience them), though they are not applicable to things as they are in themselves (*noumena*), about which no knowledge is possible. Relations such as cause and effect are the result of the organizing activity of the mind, and they are not ultimate ontological structures or essences of mind-independent noumenal reality.

Logic therefore is not etched in things in themselves. Rather, it consists of formal ordering activities imposed by the mind on everything it can experience and think. Since the laws of logic, when applied to experience, inherently structure our experience of natural objects and events (as the formal categories of experience), nature is bound to manifest a certain logical, rational order. Logic applies universally to nature as we experience it, but not to nature as it may be independent of our interaction with it.

In sum, Kant sees that we can know our world only insofar as we have partially made it by imposing on it the structures of our perception (sense intuition) and conception (thought). Kant secured the possibility of shared, communicable, objective knowledge by relativizing it to things as we structure and experience them (through our sensuous and cognitive filters). We will never know things as they are in themselves (*noumena*) independent of body and mind, but we will at least be able to know things as they appear to us (what he called *phenomena*), subject to forms of sensation and forms of conceptual organization. Even if we can never have absolute knowledge of things or events, we can know our phenomenal world because we have had a critical role in making it what it is—in constituting its essential structures. This humbles human knowledge by relativizing it to things as experienced, but in so doing, it makes knowledge possible, giving us a measure of confidence in our capacity to knowingly engage our world, thereby overcoming skepticism.

One very important corollary of this idea that knowing is a constructive activity is that subject and object (mind and world) are correlative notions. As we will articulate on neuroscientific grounds, self and world, subject and object, are co-constituted. The unity of objective experience (i.e., shared, communicable experience of objects) is a unity of the self that knows because the thing known is the result of the unifying activity of the self or subject.

> The original and necessary consciousness of the identity of the self is thus at the same time a consciousness of an equally necessary unity of the synthesis of all appearances according to concepts, that is, according to rules, which . . . determine an object for this intuition, that is, the concept of something wherein they are necessarily interconnected. (Kant 1781/1968, A108)

In simpler terms, the unity of the self is constituted in and through the unity of objects experienced by us. Or better, our self-organization is the result of our organizing of objects as known. Self and world are co-constituted and interdetermined.

This was a profound insight bequeathed by Kant to succeeding generations of philosophers. If properly interpreted, it leads to the conclusion that we are never totally alienated from our world because we are intimately involved in making it what it is, and, as such, we engage it at every step of our learning activity. In later chapters, we survey some of the scientific evidence for co-enactment or unfolding of self and world as known.

Unfortunately, Kant took his insight in a different direction, one that Dewey later made the basis of his fundamental criticism of Kantian metaphysics. Given Kant's distinction between *phenomena* (things as they appear to us under the forms of sensibility and cognition) and *noumena*, he postulated two concepts of the self. The first he called the *empirical self*—the self as an empirical (phenomenal) object among other objects in experience. This would be the experience of having a body, which involves sensations and feelings (hot or cold, wet or dry, pain or pleasure, anger or joy), proprioceptive sense of body parts and their positions in space, and a kinesthetic sense of moving our bodies. We know this self as we know other empirical objects, filtered through our forms of sensibility and brought under empirical and pure concepts. This is our self as we live and experience it. As an empirical object, Kant thought this embodied self was subject to the strict causal determinism of Newton's world. This empirical self is the phenomenal self.

Besides the self as empirical object of experiential knowledge, Kant's metaphysical and epistemological assumptions led him to posit a second— noumenal—sense of self as the source of the pure unifying activity that binds together the various elements of our experience, even though it is not itself able to be experienced. Kant took the function of understanding to be that of bringing conceptual unity to our experience. He believed that the fundamental act of mind was judgment, the synthesizing of many representations under one unifying representation (Kant 1781/1968, A68/B93), such as unifying multiple sense impressions under a concept. Mind is the innate source of this unifying, organizing activity. Kant reasoned that the ultimate unity of the self (and of objects) had to come from a transcendent ego that is the spontaneous source of order, rather than from embodied biological processes (i.e., the empirical self), which are objects receiving their organization rather than engendering it. Because Kant mistakenly thought that the sense impressions making up our ordinary experience could not come to us unified into coherent images, he resorted to the claim that the unity and identity of objects had to be the result of the activity of a pre-given (innate) form-giving faculty called understanding (or, in some contexts, reason).

But what is the "I" that unifies all of "my" representations, making both knowledge and the unity of the self possible? "It must be possible for the 'I think' to accompany all my representations; for otherwise something would be represented in me which could not be thought at all" (Kant 1781/1968, B131). He concludes that the "I" that thinks cannot be merely the empirical self because that self is the result of a prior organizing activity that makes it into an object of experience in the first place. I do not directly experience the "I" that thinks, for whenever I try to focus on my act of knowing, the "I" that thinks this thought escapes being an object of knowledge.

> All the manifold of intuition has, therefore, a necessary relation to the "I think" in the same subject in which this manifold is found. But this representation is an act of spontaneity, that is, it cannot be regarded as belonging to sensibility. I call it pure apperception, to distinguish it from empirical apperception, or, again, original apperception, because it is that self-consciousness which, while generating the representation "I think" . . . cannot itself be accompanied by any further representation. (Kant 1781/1968, B132)

Kant therefore postulated the "transcendental unity of apperception" as the alleged spontaneous source of all organizing activity by which we can have any sort of objective experience at all. The "pure" unifying activity of the self is posited, not as an empirical self that can be experienced and known, but rather on epistemic grounds as a necessary condition of our having any knowledge of objects in the first place. It is conceived as an agency of spontaneous activity that is thinkable but not knowable, since it does not show itself in experience.

The only self we can know is what Kant called "empirical consciousness," which is our knowledge of our self as an appearance, and hence as one object among others in space and time. The view of the self as a transcendent source of unifying activity made sense in Kant's mostly Christian cultural tradition, which thought of the soul as a free, disembodied, and autonomous ego. This transcendent ego, if it did exist, would be a noumenal being that is different from the causally determined empirical self, and that possesses the radical freedom necessary if we are to attribute moral responsibility. Kant's system was warmly welcomed by those hoping to find a nontheologically based rationalist ethics that kept open at least the possibility of a world behind and beyond this earthly existence, a world where the transcendent self or ego was supposedly disembodied, free, and autonomous.

Kant's account is infamously difficult and full of technical terms. So here, by way of summary, are four of his most revolutionary claims about mind and knowing:

1. We can only know things as we experience them (as *phenomena*), but not as they are in themselves (as *noumena*).

2. The unity of the self and its subjectivity is realized as the unity of objects of experience, so that subject and object, self and world, and self and knowing activity are co-constituted.

3. We can know our world because we have made it in our own image (i.e., through organization we impose, via structures of perception and thought, upon any object of possible experience).

4. The noumenal self is a transcendent source of spontaneous activity that gives form to everything we can experience, but that cannot itself ever be made an object of experience or empirical knowledge.

It is our contention that each of these key tenets of Kant's view of self and knowledge harbors an important insight about the nature of knowing, although teasing out what is profound and correct in his view requires discarding certain erroneous ontological assumptions that led Kant to postulate a world and a self beyond nature. We first describe what we take to be Kant's greatest contributions, and then argue that each of them has to be tempered and reinterpreted (sometimes substantially) in light of current science.

First, Kant was right that we know things only as we experience them, filtered through the structures of our perception, action, feeling, and cognition, which give order to and make sense of what we experience. This does not, however, justify the postulation of some unknowable otherworldly noumenal realm hidden behind nature and selfhood. The idea of a noumenal realm that we can neither experience nor know needs to be replaced with a process metaphysics that reveals how and why there is always more to any object or event than we can possibly experience or know at any given time. We also need a cognitive neuroscience that can explain why the ultimate processes of our selfhood and experiences operate mostly beneath our conscious awareness. Nature is not a hidden world, beneath or beyond experience, but rather nature *is* experience—both what is experienced and how it is experienced. Experience is our only way of penetrating the depths of natural events, and it provides the means for us to be in touch with nature, although never in a totalizing, comprehensive, or final way. Because of possible differences in their bodies, purposes, and values, different animals will encounter different affordances from the "same" environment. So, our universe supports a plurality of possible understandings and ways of inhabiting nature. That ontological and epistemic pluralism, in which no conception can ever exhaust the depth and richness of any experience, is the proper ontological and epistemological payoff of Kant's flawed conception of the noumenal. The notion of some noumenal reality forever hidden from us needs to be replaced by an awareness of the inexhaustible depths of a natural world in process.

Kant's second insight—that the unity of the self is realized in and as the unity of objects experienced so that self and world are co-constituted—is borne out by the biological and neuroscience research that we will explore. Kant saw one important implication of this intertwining of self and object, namely that, contra the skeptics, we really can have empirical knowledge of things and events. As co-constituted, both self and its objects are in and of nature, and so as knowers we are in touch, through organism-environment interactions, with how things are or can be. Unfortunately, Kant's phenomenal/noumenal dualism leads him to claim that we can never know ourselves as we "really" are, which puts the pure self in a realm beyond nature, forever closed off to our experience. If, following Dewey, we conceive of nature as what there is, then the hypothesis of an unknowable noumenal realm behind the natural world is an unnecessary appurtenance that does more harm than good. It is neither scientifically defensible nor necessary to account for the depths and richness of experience and knowledge.

It is important to be clear that it does *not* follow from what we have said that we humans can ever have absolute, comprehensive, or final knowledge of anything. Nature is an ever-developing creative process, and so none of our intellectual selections of significant patterns, concepts, or values could ever possibly tell the "whole truth and nothing but the truth." But that is no disparagement of the genuine knowledge of our world that we can have, because we are in and of that world and not separated from it.

Third, Kant helped us see that we can know our world mostly because we have co-created it. That shared world is shaped by the structure of our perceptual mechanisms, by our conceptual selections as a basis for thinking and reasoning, and by our purposes and values that determine what we care about and focus on as significant and relevant. This view is well supported by cognitive neuroscience evidence (examined in chapters 7–10), which emphasizes both the objective aspects of experience and also the subjective meaning of it for our lives. That we have co-created our world does not mean that the world is what we have conceived it as being. We must never forget the *co-* in co-creation. Our perceptual and conceptual selections and prejudices about what matters are only part of what makes nature. We all know, or should know, that nature "talks back" to us all the time, refusing our clunky, inapt, and reductive conceptualizations of it. Dewey wrote entire books to remind us that "knowing" or "cognitive" experience is most definitely not the totality of experience, which exceeds any conceptual or propositional patterns we project onto it.

This appreciation of the limits of human cognition must be a fundamental insight for building a natural philosophy. What Dewey called the "philosophical fallacy" is the selecting of certain patterns or dimensions that suit our purposes for inquiry, and then treating them as if they constituted reality itself, or at least all that matters. The dangerous idea here is that nature is only what we conceptualize or know it to be. The moral here is that there is much more to experience than what we know of it, and failure to include these other relevant dimensions makes our thought ill-adapted to our situation.

Kant's fourth claim, that there is a transcendent ego behind experience that is the source of all organizing activity, is more mistaken than insightful. From a scientific point of view, there is no need to posit a pure self, nor is there scientific evidence for any such reality. Kant's idealistic metaphysics stem from the classic idealist assumption that physical nature is mechanistic and deterministic, so that there must be some ontologically different kind of agency that is the spontaneous source of activity. Once you see that activity is characteristic of all living things, there is no need to posit a noumenal realm of spontaneity and freedom in order to underwrite the possibility of activity. The challenge from a naturalistic perspective, of the sort we develop here, is to explain all the marvelous phenomena of life and mind without reference to supernatural causes, forces, or entities. Obviously, this is a monumental task, but the absence of adequate explanations of certain phenomena should provide no justification for hypothesizing transcendent or nonnatural phenomena.

Even though Kant's fourth tenet is mistaken, and badly so, there is an important truth in the way he recognizes that there is more to our experience and cognition than we can know. Perhaps he is right when he observes that the "I" of the "I think" cannot be known at the moment it is thinking. But there is a very different, and less problematic, way of interpreting this truth than projecting a transcendental unity of apperception operating in a noumenal realm beyond experience! We suggest that the proper interpretation of these limitations of our knowing is the cognitive unconscious (Lakoff & Johnson 1999) and the monumental role of nonconscious processes in our experience (Damasio 1999; Edelman & Tononi 2000; Tucker & Luu 2012). The recognition that the vast majority of our experience and cognition goes on unconsciously does not justify Kant's postulating of a transcendent ego hidden forever behind or beyond experience in a mysterious noumenal realm. Instead, it reminds us that the functional operations of perception, body movement, organism maintenance, meaning making, understanding, and reasoning happen mostly by unconscious mechanisms and processes that we cannot be aware of at the same time we experience what they enact. This is one reason why reflective phenomenological description can never tell us the whole story of cognition and action. We need the various methods of several sciences to explore the working of these nonconscious operations. We need biology, physiology, neuroscience, computational modeling, cognitive linguistics, and multiple psychological and philosophical orientations to plumb the hidden depths of experience.

3.5   Toward a Naturalistic Philosophy of Mind and Knowing

The upshot of this chapter is that self and world are co-constituted realities that are always in process. The learning you do throughout your life emerges from your self as presently formed as a result of your prior experience, and it issues in the dual forms of a reconstructed self and a different meaning of the world. We will later investigate the neural basis for how this ongoing reconstruction of the self operates. The biggest obstacles to such a naturalistic conception of mind are the metaphysical dualisms of mind versus body, *phenomena* versus *noumena*, cognition versus emotion, inner versus outer, and theoretical versus practical knowledge. Therefore, in the next chapter we want to turn to the richest and most scientifically informed philosophical orientation we know of that helps us overcome these and other dualisms and to recognize a view of mind as an embodied emergent functional process. That nondualistic, nonreductivist, and naturalistic philosophical perspective is pragmatism.

4 A Pragmatist Naturalistic Framework for Embodied Mind and Knowing

The account of mind and knowing that we develop here draws heavily on research in biology, psychology, cognitive science, and neuroscience. We cannot give an adequate account of knowing only by focusing on specific bodies of scientific research. We need an interdisciplinary investigation into the nature of mind, experience, meaning, thought, feelings, emotions, motivation, and values in order to see how all of these fit together to give rise to knowing. Knowing is a whole-body process of engagement with our world. It is not a single modular function, nor even a combination of multiple modular functions. Various scientific approaches and methods, with their selective focus on particular structures, processes, and functions, often do not, or cannot, provide the larger picture in which we can see how all of these key components blend together to make knowing possible. That is where philosophy comes in—to put our acts of knowing in their broader experiential context and to explore the meaning of knowing for our lives.

Philosophy is not a competitor with the sciences for empirical truths about mind and world. Philosophy's role is (1) to examine critically the assumptions, methods, and values underlying the various sciences of mind; (2) to construct a comprehensive view of how these scientific results can be integrated with humanistic and aesthetic perspectives; (3) to look for converging evidence from these different methods and practices, which provides our best chance for a biologically and psychologically adequate theory; and (4) to explore the implications of this research for our lives—for who we are and how we ought to live.

We believe that the most sophisticated, nuanced, and scientifically responsible philosophical orientation capable of carrying out these important reflective tasks is Pragmatism. Pragmatism emerged in America in the latter part of the nineteenth century and came to fruition in the first four decades of the twentieth century. It suffered an eclipse by mid-century, with the onslaught of Anglo-American analytic philosophy, but it has undergone a remarkable revival over the past twenty years, often in conjunction with the rapid growth of the cognitive sciences (Johnson 2007; Madzia & Jung 2016; Solymosi & Shook 2014). It is now possible to see how mind science works together with pragmatist philosophy to develop the basis for a more adequate natural philosophy of mind and knowing.

We adopt a Pragmatist approach for three reasons. First, Pragmatism strives to be empirically responsible, by which we mean that it realizes that it can coevolve through critical dialogue with the best scientific research on mind, thought, language, and values. It eschews armchair metaphysical speculation in favor of an appreciation of the need for an ongoing constructive engagement with the sciences, balanced by recognition of the partiality and limitations of any particular scientific approach. Second, Pragmatism appreciates that in order to understand the *experience* of knowing, we have to start with a rich and nonreductive account of experience, of the sort one finds in the remarkable phenomenological descriptions provided by William James, John Dewey, and other major pragmatist philosophers. Third, Pragmatism is motivated by the commitment that any philosophy worth having ought to help us live more meaningful, more intelligent, and more fulfilled lives. Such a philosophy has existential and moral bite, so to speak. With respect to our focus on knowing, then, a pragmatist perspective is important, not only because it helps to construct a scientifically adequate theory of knowing, but also because it tells us something significant about the meaning—the existential import—of knowing in our lives.

We propose to focus primarily on John Dewey's theory, although readers will see commonalities to William James, Charles Sanders Peirce, and other pragmatists. We have selected Dewey because we find his treatment of the key components of a theory of knowing to be remarkably similar to research coming from contemporary mind science that at once supports, extends, and enriches Dewey's general framework (Johnson 2017; Solymosi & Shook 2014). Other philosophical orientations, such as phenomenology (Gallagher 2005; Merleau-Ponty 1962) have much to contribute to this philosophy-science dialogue, but Pragmatism, in both its classical and contemporary incarnations, has a more sustained and expansive engagement with the sciences of mind than any other philosophical perspective.

By focusing mostly on Dewey's version of pragmatism, we open ourselves to the charge of partiality and of failing to muster the full resources of a pragmatist theory of mind. Obviously, Dewey does not represent all of pragmatism, and other authors (most notably C. S. Peirce, William James, and others such as Frank Ramsey, Alain Locke, and George Herbert Mead) made significant contributions concerning the key issues we are addressing. Incorporating these other pragmatist resources would have substantially enriched our account, but it would also have substantially lengthened that account in a book that some may think is already quite long enough. So, we have stuck primarily with Dewey, making only occasional reference to other pragmatist-oriented philosophers, with the assumption that champions of these other thinkers could easily muster resources from their favored figure that might clarify, expand upon, and even correct our account of pragmatism.

What follows in this chapter are a number of key tenets of what we take to be a Pragmatist theory of mind and knowing. We provide a general framework for reconceiving mind as an embodied social process—and knowing as an ongoing embodied activity, driven by emotions and motive controls—that transforms our understanding by deepening and expanding the meaning of our present situation.

4.1 Organism-Environment Interaction as the Source of Experience

For living animals, everything begins and ends with the arc of organism-environment interaction that constitutes experience.

> The first great consideration is that life goes on in an environment: not merely in it but because of it, through interaction with it. No creature lives merely under its skin; its subcutaneous organs are means of connection with what lies beyond its bodily frame, and to which, in order to live, it must adjust itself, by accommodation and defense but also by conquest. At every moment, the living creature is exposed to dangers from its surroundings, and at every moment, it must draw upon something in its surroundings to satisfy its needs. The career and destiny of a living being are bound up with its interchanges with its environment, not externally but in the most intimate way. (Dewey 1934/1987, 19)

This intimate ongoing transaction between organism and environment is what Dewey called *experience*. He challenges the tendency to regard experience merely as the process of subjectively experiencing something. Instead, experience is both *how* things are experienced (i.e., the subjective dimension) and equally *what* is experienced (i.e., the objective dimension). Dewey eschews mind-body dualism by recognizing that mind, feeling, thought, and knowing are emergent levels of organism-environment interaction rooted in the experiential endeavors of a human creature engaged in a continual quest for survival and enhancement of well-being. Therefore, it is crucial to start with a rich, nonreductive notion of experience if we want to avoid the perennial mind-body dualism that plagues so many philosophical accounts of knowledge. If you think of experience as merely subjective—as experiencing within one's mind—and if you think of what is experienced as out there in the mind-independent world, then you create a metaphysical and epistemic gap between the inner and the outer, mind and world. Once such a radical dualism is enshrined, you will never be able to explain how the inner (thought, feeling) could ever connect up with, or map onto, the outer (objects, events in the world) because you already began by assuming that mind and world are essentially different kinds or categories of being. This typically leads to the traditional problem of knowledge, which becomes how to bridge the gap between subjectivity and objectivity, between what is "in the mind" and what is "out there in the world."

Dewey saw that the only way to avoid this skeptical problem of how to bridge the mental versus physical gap was to realize that there is no such metaphysical or epistemic gap in the first place. We never were fundamentally divorced from the world because everything we are and do emerges from our transactions with our surroundings. Consequently, even though we can be mistaken in our expectations about our world, we are always *in* and *of* the world—engaging our surroundings, albeit contingently and fallibly. In trying to explain mind, there is thus no need to conjure up a separate metaphysical realm of transcendent objects, forces, causal sources, or qualities, since we can explain natural events in terms of natural relations, forces, qualities, and interactions. Dewey's brand of realism rests on a nondualistic account of experience in which "subject" and "object" are correlated aspects of the continuous interaction of organism and environment. There is no chasm between the experiencing self and nature because experience is our mode of access to nature—our way of inhabiting, exploring, and acting in and through nature.

> Experience is of as well as in nature. It is not experience which is experienced, but nature—stones, plants, animals, diseases, health, temperature, electricity, and so on. Things interacting in certain ways are experience; they are what is

experienced. Linked in certain other ways with another natural object—the human organism—they are how things are experienced as well. Experience thus reaches down into nature; it has depth. (Dewey 1925/1981, 12–13)

If we are always *in* and *of* nature, then there is no great mystery about how we can knowingly engage our world. Experience is not an epistemic veil that blocks our direct access to natural goings-on; instead, it is our only satisfactory access to the workings of nature, and it makes knowledge possible.

4.2   Life and Homeostasis within an Organism in Its Environment

The life of any organism is an ongoing struggle to maintain its integrity as a living creature. As neuroscientist Antonio Damasio explains: "Life is carried out inside a boundary that defines a body. Life and the life urge exist inside a boundary, the selectively permeable wall that separates the internal environment from the external environment. . . . If there is no boundary, there is no body, and if there is no body, there is no organism" (Damasio 1999, 137).

The key to survival and enhanced well-being of any organism is homeostasis—the preservation and/or restoration of the dynamic equilibrium of the internal milieu that is necessary for the functioning of the organism. The traditional conception of homeostasis involves maintenance of a predetermined set-point equilibrium, such as when a mechanical thermostat kicks on a heater whenever the ambient temperature drops below a preset level, causing the system to produce heat until the desired temperature is restored. In living organisms, homeostasis governs myriad bodily systems, such as those that regulate temperature, pH levels, oxygen saturation, salt levels, and so on in order to maintain the body's internal milieu within a certain narrow range that is necessary to sustain life functions. However, because merely responding to deviations from a preestablished set point is often not enough to insure successful functioning for most animals, the term "allostasis" has been coined to denote the more dynamic process of constituting a new equilibrium in response to new conditions that require the organism to reorganize itself as a living system (Schulkin 2011, 133–136; Tucker & Luu 2012, 70).

> Allostasis emphasizes regulation that is an adaptation to change; not just in reaction to it, but in anticipation of it. Unpredictable events are a constant feature of the life cycle for most animals, and the need for stability and consistency are a constant characteristic of most animals. Within both the physiological and social domains, allostasis is a means of achieving stability in the face of unpredictability. (Schulkin 2011, 6–7)

Damasio tends to use the term "homeostasis" to include allostasis, but we will use both terms in order to distinguish the mostly restorative processes from the more dynamic anticipatory and transformative processes.

4.3   Knowing Is Based on Values of the Organism

Homeostasis is a fundamental value for all organisms, and therefore components of a situation that are required for life maintenance are subsidiary values.

> My hypothesis is that objects and processes we confront in our daily lives acquire their assigned value by reference to this primitive of naturally selected organism values. The values that humans attribute to objects and activities would bear some relation, no matter how indirect or remote, to the two following conditions: first, the general maintenance of living tissue within the homeostatic range suitable to its current context; second, the particular regulation required for the process to operate within the sector of the homeostatic range associated with well-being relative to the current context. (Damasio 2010, 49)

The key point here is that all organism-environment interactions involve values. Acts of knowing occur as part of our attempt to engage our environment in ways that help us survive and realize well-being. Consequently, far from being the value-neutral, nonemotional activity proposed by traditional theories of knowledge, knowing is, at its roots, profoundly shaped by our values, motives, and emotions. Knowing is a doing, an activity, and, insofar as it is part of our loosely purposive activity in the world, it is our most intelligent means for realizing values dear to us. Knowledge is not always directly related to our immediate survival, but it is nonetheless grounded in the biological values of the organism, the chief of which is homeostasis/allostasis. At higher levels of organization, such as interpersonal relations and membership in larger communities, knowing serves additional values, such as interpersonal trust, social bonding, social justice, and cooperative group activity, but it is shaped by values at every level of experience.

4.4   Knowing in the Context of Need-Search-Satisfaction

The life of animals manifests a specific recurring pattern in which the organism falls in and out of harmony with its surroundings. Dewey recognized that this pattern of behavior plays a crucial role in our experience, self-identity, and modes of knowing.

The habits we develop are not just internal dispositions; rather, they are the result of organic responses to conditions in our environment. Habits incorporate aspects of our environment, and they are tuned up in intimate ongoing relation to environmental conditions. Dewey reminds us that there would be no habits without an environment and that our surroundings structure the character of our habits:

> Habits are like functions in many respects, and especially in requiring the cooperation of organism and environment. Breathing is an affair of the air as truly as of the lungs; digesting an affair of food as truly as of tissues of stomach. Seeing involves light just as certainly as it does the eye and optic nerve. Walking implicates the ground as well as the legs; speech demands physical air and human companionship and audience as well as vocal organs . . . They are things done by the environment by means of organic structures or acquired dispositions. (Dewey 1922/1988, 15)

We can think of ourselves as complexes of interpenetrating habits that incorporate and are shaped by the affordances for interaction provided by our surroundings. Through most of our daily routines, we act habitually and unreflectively, going about our business of living in the habit channels established by our prior experience, which forms our current self-identity. Sometimes, however, and more often than we might think, new conditions arise that call into question a certain habit or cluster of them. Instead of moving fluidly and unthinkingly forward, we are taken aback by the indeterminacy of our situation, where we are no longer certain how to act. Our habits have fallen out of sync with our surroundings. Dewey called this indeterminacy *need.*

Need puts us in a situation where we cannot move forward as usual because some of our prior habits have become ill-suited to our present situation. However, need can have a positive, motivating role insofar as it initiates a process of bodily inquiry that Dewey calls, alternately, "demand," "effort," and "search." This arrest and frustration motivates us to search for a way to reconstruct our situation so as to reduce the indeterminacy and satisfy the need that characterizes our current state of affairs. In later chapters on motivational controls, we explore the neurochemical basis of this need-search process.

Whenever we are successful in our efforts to reestablish a working equilibrium, we speak of *satisfying* our needs. The conflict of habits and urges is resolved, at least to such an extent that we don't feel completely blocked, and we can "move forward" for the moment. Dewey summarizes this fundamental experiential pattern of need-effort-satisfaction as follows:

By need is meant a condition of tensional distribution of energies such that the body is in a condition of uneasy or unstable equilibrium. By demand or effort is meant the fact that this state is manifested in movements which modify environing bodies in ways which react upon the body, so that its characteristic pattern of active equilibrium is restored. By satisfaction is meant this recovery of equilibrium patterns, consequent upon the changes of environment due to interactions with the active demands of the organism. (Dewey 1925/1981, 194)

This structure of biologically based transformative activity emerges when needs arise, the satisfaction of which is not possible under the impress of our current habits of response and behavior. Need is a condition in which the outcome of a developing experience is indeterminate and the organism is in a state of tension and relative instability. Search involves effort to reduce the indeterminacy of the situation. Satisfaction is either a return to temporary relative stability (homeostasis), or the establishing of a new dynamic equilibrium (allostasis), which enables some measure of functioning for the organism.

To illustrate this fundamental pattern of animal life, consider the following three manifestations of body-based need-search-satisfaction at the level of organism-environment interaction. Case 1: You haven't eaten in a while and you become hungry (need). You start to feel anxious and begin to scrounge around for some grub (search), and then you find food to end the craving (satisfaction). Case 2: You've been sitting too long and feel cramped and uncomfortable (need). You stand up, stretch, and move around the room (search/effort). You feel more relaxed and invigorated, and can return to your work (satisfaction). Case 3: You realize that your partner is upset with you about something. You don't know what, the whole situation is tense, and you are ill at ease (need). You ask them if something is bothering them and whether there is anything you can do for them (search/effort). As the two of you talk, you feel the tension releasing. A measure of harmony is reestablished, at least for the present (satisfaction).

The first two cases above are mostly automatic and unconscious animal responses, while the third involves social interactions and often has a more conscious reflective dimension. *Dewey argues that even our most imaginative and highly reflective episodes of theoretical inquiry and pursuit of knowledge manifest the same need-search-satisfaction pattern that characterizes all animal life.* One of Dewey's most significant insights about knowing is that this animal need-search-satisfaction pattern is not merely a structure of nonreflective biological functioning to maintain allostasis. It is that, but it is also much more. In creatures like us, who have the capacity to experience meaning and to think, this very same threefold process operates at the level of knowledge processes involving abstract conceptualization and reasoning. As we will see in chapter 6, humans recruit body-based structures (for perception, motor activities, and feeling) operating at the level of our biological functioning, repurposing them (evolutionarily through exaptation) for so-called higher cognitive acts, such as knowing, planning, and reasoning.

Dewey was prescient in recognizing the inescapably bodily sources of meaning and thought that have recently become the darlings of what is known as embodied cognition theory (see chapter 6). Consequently, the need-search-satisfaction process applies as much to abstract reasoning, reflective problem solving, and knowing as it does to successful biological and bodily functions. This is one of the most profound ways in which knowing is embodied—continuity between sensory, motor, and affective processes and higher, more reflective, cognitive processes.

On this view, to know something is to grasp its meaning for experience in such a way that you become better able to operate within your world. The need for knowledge results from the indeterminacy of your present situation. Intelligent inquiry is the search process, by which you explore the meaning of your situation, look for relations among things, and trace out probable consequences connected with a thing or event. Satisfaction is feeling a release of energies into the world as a result of your knowing activities, so that you can move forward intelligently rather than blindly. Knowing is not a fixed state, but rather an ongoing construction of relations with your surroundings, by means of which you are able to be more or less "at home" in your world.

At the level of reflective thinking, intelligent inquiry is knowing activity geared toward helping us reestablish a dynamic equilibrium that has been disrupted by encounter with an indeterminate situation. Knowing activity is about discerning the meaning of our situation in a way that frees us, to some degree, from the indeterminacy that blocks our action. This opening up of new possibilities enables us to grasp a richer and more complex meaning of our situation in order to adapt to our surrounding conditions and to act effectively in transforming them. In subsequent chapters we will explain the neural basis for the subjective dimensions of this need-search-satisfaction process.

The processes of knowing are entirely natural. They arise through the growth of our bodies and brains, shaped by both our evolutionary history and our individual development. They work from the bottom up, based on emergent sensory, motor, and affective processes that are recruited for "higher" cognitive operations, to be described in chapter 6. They are not brought into (or onto) experience from outside, that is, from some alleged structure of pure reason or from some preexisting realm of essences. That would only reinstate a gap between mind and nature. Instead, operations of thinking and knowing emerge from our most basic bodily ways of surviving and flourishing in our changing environment. "The organs, instrumentalities, and operations of knowing are inside nature, not outside. Hence, they are changes of what previously existed: the object of knowledge is a constructed, existentially produced, object" (Dewey 1929/1984, 168). Knowing is thus a way of transforming experience, not a way of re-presenting it. From this "embodied" perspective, there is no traditional problem of knowledge—no problem of how mind can possibly know what is not mind—since we give up the dualistic metaphysics that separates mind from body and world. We see the bodily organism in its environment as the locus of meaning, thought, and knowledge. Such a view "installs man, thinking man, within nature" (Dewey 1929/1984, 168). It is in this sense that knowing activity is a profoundly embodied, situationally embedded, and dynamically enactive process.

4.5 The Social Constitution of Mind

The obvious problem for any naturalistic, embodied theory of knowing is to explain how we get from basic bodily transactions with our surroundings to our highest acts of abstract conceptualization and reasoning without relying on some transcendent source. In Dewey's framework, that amounts to explaining how we move from organic biological need-search-satisfaction to this same three-fold pattern in our more reflective and abstract activities of knowing.

The key to the continuity between bodily (biological) need-search-satisfaction and the same structure in abstract conceptualization and reasoning is an emergence theory of mind. Mind, meaning, thought, reason, language, and intelligence are emergent functions that become possible through the increasing complexity of organisms and their environments, through both evolutionary history and more immediately in each individual's development. The need-search-satisfaction process operates unconsciously in lower animals and also in humans most of the time. However, in creatures like us—creatures capable of meaning and communication through symbolic interaction—that tripartite process can be executed consciously too. In other words, it can become a process by which we explore the meaning of objects and events, and then use this grasp of meaning as a basis for acting in the world in a generally purposive manner.

This naturalistic approach rests on an account of how mind, and our capacities for meaning making and thought, can arise from our sensory, motor, and affective processes. Dewey described three major "plateaus" of emergence, each with its own distinctive functional organization: the physical, the psychophysical, and mind. The *physical* plateau is the level of inanimate material causal interactions. The *psychophysical* realm is the locus of life processes in organic beings, by virtue of which they can strive, mostly unconsciously, to maintain the allostatic conditions of life within their organism boundary. Dewey observes that, already at this level, there are values operative, since living organisms are preferentially directed toward maintaining certain bodily states and appropriating aspects of their environments that are necessary for survival and growth. At this second, psychophysical level, there is sensitivity toward certain environmental affordances, and some animals have evolved capacities for feeling different qualities of things and situations. Need-search-satisfaction operates at this second level mostly unconsciously in order to ensure the survival and well-being of the animal. Fortunately, we don't have to activate every breath we take or every beat of our heart consciously in order to maintain appropriate levels of oxygen in our body and brain. Nature has automated that function to focus our attention elsewhere.

It is the third plateau—that of *mind*—that makes possible more reflective knowing. Dewey notes that in the psychophysical realm, an animal might feel a quality of some perceived object or event, but it would not know that quality. "Complex and active animals have, therefore, feelings which vary abundantly in quality, corresponding to distinctive directions and phases . . . of activities, bound up with distinctive connections with environmental affairs. They have them, but they do not know they have them. Activity is psychophysical, but not 'mental,' that is, not aware of meanings" (Dewey 1925/1981, 198). To have this higher form of knowledge requires the capacity to explore the meaning of things in all of their complex interrelations. Consequently, mind is an intersubjective process in which individuals, interacting in communities through shared meanings, coordinate their activities.

This social constitution of mind and self reveals the fact that knowing is not just an inner act of an individual person, but instead depends on our participation within communities of inquirers and actors via systems of meanings (e.g., concepts, practices, institutions, rituals). Dewey argues that the move from the plateau of the psychophysical (life) to the plateau of mind results from the capacity for meaning making that comes with the introduction of language as a means of social cooperative activity. "'Mind' is an added property assumed by a feeling creature, when it reaches that organized interaction with other creatures which is language, communication. Then the qualities of feeling become significant of objective differences in external things and of episodes past and to come" (Dewey 1925/1981, 198). Mind is *not* an entity, structure, or process that exists only in an individual brain or body. Mind *is* an interpersonal, socially constituted, and culturally shaped system of possible meanings. "Mind denotes the whole system of meanings as they are embodied in the workings of organic life; consciousness in a being with language denotes awareness or perception of meanings. . . . The field of mind—of operative meanings—is enormously wider than that of consciousness. Mind is contextual and persistent; consciousness is focal and transitive" (230).

Dewey's interpersonal, social, and cultural conception of mind highlights our participation with others in developing systems of meaning and thought. Mind is not a possession of an individual person, but rather a system of possible meanings (through shared language and cooperative practices) in which each individual person can participate, thereby drawing on vast resources for meaning and thought. Figure 4.1 captures the interpersonal, shared, and always developing character of mind.

The traditional conception posits a developed individual self, endowed with mind, and conceived as a "formal capacity of apprehension, devising and belief" (Dewey 1925/1981, 169). One could have such a mind without reference to others. By contrast, Dewey's self is defined by its participation in the culturally evolving system of meanings and practices (i.e., mind).

> But the whole history of science, art and morals proves that the mind that appears in individuals is not as such individual mind. The former is in itself a system of belief, recognitions, and ignorances, of acceptances and rejections, of expectancies and appraisals of meanings which have been instituted under the influence of custom and tradition. (Dewey 1925/1981, 170)

What things mean to us, what we think and know, and how we think and know all depends on social relations and evolving cultural institutions,

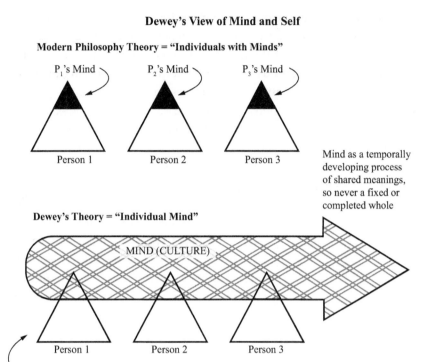

Figure 4.1
Dewey's model of self and culture.

values, and practices that make us who and what we are. *Mind and self are inescapably social and cultural, even as they are biological.*

The epistemological implications of this view of mind and selfhood are profound and far-reaching. Reflective knowing is based on our ability to discern the meanings of objects and events we encounter, thereby enabling action that transforms both self and world. Insofar as new meanings are discovered and made, we then are a new self, dwelling in a different world than we inhabited at the outset of our need-search-satisfaction adventure. Plateaus of mind emerge when we can interact with other people to communicate the meaning of things, which is to say, to trace out, via symbolic interaction, the sources and consequences of things or events in experience. According to the pragmatist conception, meaning is the past, present, and future (possible) connections and relations to something we experience. Some of these experienced relations are about what has come before in the history of the object or event, and some are projected future consequences of the object or event in its relations to other things, so meaning has both a backward-looking and an anticipatory, projective dimension.

Dewey thought that meaning, as public and shareable, depends on a shared system of signs and interpersonal transactions using symbols. He called this "language," but he had a broad sense of language, including a wide variety of symbolic interactions in art, music, dance, technology, gesture, and ritual practices (Dewey 1934/1987, 51–52). Dewey sums up his view of the emergence of mind as the means whereby humans become capable of meaning and thought necessary for knowledge.

> As life is a character of events in a peculiar condition of organization, and "feeling" is a quality of life-forms marked by complexly mobile and discriminating responses, so "mind" is an added property assumed by a feeling creature, when it reaches that organized interaction with other living creatures which is language, communication. Then the qualities of feeling become significant of objective differences in external things and of episodes past and to come. This state of things in which qualitatively different feelings are not just had but are significant of objective differences, is mind. Feelings are no longer just felt. They have and they make sense. (Dewey 1925/1981, 198)

We will soon explore the neural basis for this developmental landmark of moving from life to mind, but, for now, the key point is that the capacity for acquiring and communicating meaning is the basis for our highest human achievements of mind. This raises the question of whether it is legitimate to attribute mind and knowing to animals that do not seem to possess language and the meaning language makes possible (McKenna 2018).

On the one hand, it seems many animals have some kind of knowledge insofar as they can nonreflectively discriminate qualities in their environment and respond appropriately to them to further their own well-being. On the other hand, they do not grasp the meaning of things in the same way, or with the same depth, as do creatures with language systems. Still, this appears to be more a continuum than a discrete break between humans and other animal species. We suggest recognizing a continuity, rather than a gap, between nonreflective bodily adjustments to environmental changes and language-based acts of reflective knowing. Knowing runs the gamut from nonconscious activities by which an organism adjusts to its environment, to our highest human acts of reflective inquiry into the nature of things. Knowing is about grasping the meaning of one's situation, and language vastly enriches and extends our ability to explore the meaning of things, so it plays an important role in most of our knowing. However, there are activities of knowing that do not directly require language in the sense of linguistic communication. The one constant feature at both ends of this continuum is that they involve grasping the meaning of what is happening. If, like Dewey, you want to reserve the terms "mind" and "knowing" for creatures with language, it is nevertheless important to acknowledge the continuity between organism-environment adjustment activities and more abstract and reflective knowing activities.

Even Dewey, after saying that meaning depends on language, goes on to recognize a certain grasp of meaning that is not necessarily linguistic. He calls that activity *making sense* of things, and he contrasts *sense* with *signification*, the use of signs (in a language) to communicate meaning.

> Sense is distinct from feeling, for it has a recognized reference; it is the qualitative characteristic of something, not just a submerged unidentified quality or tone. Sense, is also different from signification. The latter involves use of a quality as a sign or index of something else, as when the red of a light signifies danger, and the need of bringing a moving locomotive to a stop. The sense of a thing, on the other hand, is an immediate and immanent meaning; it is meaning which is itself felt or directly had. When we are baffled by perplexing conditions, and finally hit upon a clew [sic], and everything falls into place, the whole thing suddenly, as we say, "makes sense." . . . The meaning of the whole situation as apprehended is sense. . . . Whenever a situation has this double function of meaning, namely signification and sense, mind, intellect is definitely present. (Dewey 1925/1981, 200)

Signification is the easier of the two meaning functions to understand. It is the use of something as a sign pointing to some experience—to antecedent conditions, to relations, and to consequences in experience. Smoke (a sign) has meaning through its connections to other experiences. Its meaning is connected with fire, heat, a rising cloud of smoke, coughing, burning eyes, cleansing the spirit in an indigenous ritual ceremony, calming a colony of bees, and much more, depending on the context. This is the classic pragmatist conception of meaning as implications of a thing or event for experience and action.

The notion of sense is more difficult to comprehend, but equally important to signification, because it constitutes our grasp of the immanent meaning of the situation within which and in relation to which our thinking arises. And if knowledge is going to be worthwhile, it must be relevant to our situation. In his 1930 essay "Qualitative Thought," Dewey does not use the term "sense," but instead refers to the pervasive qualitative unity that defines an experienced situation:

> What is intended may be indicated by drawing a distinction between something called a "situation" and something termed an "object." By the term situation in this connection is signified the fact that the subject-matter ultimately referred to in existential propositions is a complex existence that is held together in spite of its internal complexity by the fact that it is dominated and characterized throughout by a single quality. By "object" is meant some element in the complex whole that is defined in abstraction from the whole of which it is a distinction. The special point made is that the selective determination and relations of objects in thought is controlled by reference to a situation—to that which is constituted by a pervasive and internally integrating quality. (Dewey 1930/1981, 246)

Dewey says our thinking and knowing activities are grounded in and related to the situation in which we find ourselves, but only when we orient our thinking to what is relevant in defining and characterizing the situation. We have to be able to "make sense" of the relevant context of our particular acts of thinking and knowing. Otherwise, our thought would not be appropriately connected to what we purport to be thinking about. Dewey concludes that "the underlying unity of qualitativeness regulates pertinence and relevancy and force of every distinction and relation; it guides selection and rejection and the manner of utilization of all explicit terms" (Dewey 1930/1981, 24–48). Dewey's language here is somewhat awkward, but his point is critical. *The starting point of all properly grounded thinking and knowing must be the felt unity of a whole situation that is beginning to make sense to us, in a way that gives rise to further exploration through signification.* (For a detailed analysis of the role of felt qualitative unity as the basis for thinking, see Gregory Pappas 2016, 20). In chapters 7–9 we will explore the neurophysiological basis for our ability to make sense of a situation.

We cannot describe this qualitative unity that gives meaning to the context of our inquiries and actions because any description would involve only selected qualities, thereby missing the unity of the whole situation. But what we can have is a felt sense of the situation as generating relevant distinctions that make it possible for us to inquire further into the meaning of what we are experiencing. William James (1890/1950) referred to this as the "fringe" or "penumbra" of meaning that surrounds any particular focus of meaning. Later, in chapters 9 and 10, we suggest that this feeling has its basis and motivation in the way our brains process our transactions with our environment, especially through mechanisms of motive control. Our neural account will explain how knowledge includes both our sense of feeling grounded and oriented in a situation and also our ability, through signification, to anticipate how our situation will develop in relation to past events and future consequences.

We have devoted considerable attention to the relation between sense and signification because they represent two critical modes of knowing, or perhaps stages in the knowing process (to be developed in neural terms in chapters 7–9). Moreover, while signification requires the use of signs and depends on language, the initial process of making sense does not necessarily rely on language. Prelinguistic children begin to grasp the meaning of objects, events, actions, and situations. Children learn that if they don't hold on tightly to the monkey bars, they will fall. They learn that if they take their playmate's toy, the playmate will be angry. They learn that if they put their hand in those colorful flames, they could get an "oweee!" They learn the meaning of things in their world. Of course, when they acquire language, the richness of meaning available to them grows exponentially, even though they were set on this path of meaning prior to language. Moreover, the nonlinguistic arts—painting, sculpture, music, dance, architecture, and so on—richly enact meanings that are not linguistically based, but rather rooted in our primal bodily engagement with our world (Johnson 2007, 2018).

With respect to knowledge, Dewey recognized two ways in which a living animal can be knowingly "at home" in its world. The first involves grasping the sense of one's situation (i.e., a holistic, feeling-infused take on one's context), which is essential to orienting oneself meaningfully, as a basis for further inquiry. The second is a reflective mode of inquiry that uses signs to explore the meaning of a situation, and more consciously enacts the need-search-satisfaction process to test out in experience our expectancies and hypotheses about how things will develop under certain conditions. Via this second process, our knowledge becomes subject to experiential and experimental testing, both in the form of seeing whether anticipated experiential outcomes actually transpire in everyday life and in the form of more formal and controlled experimental testing in the sciences.

4.6 Knowing as a Process of Learning the Meaning of Things

Dewey urges us to see knowing as a process of inquiry into the meaning of things. To understand the meaning of a thing or event is to gain a sense of how it is related to other things and events—both what has preceded it and what its anticipated experiential consequences are. The meaning of a thing is never fixed or final, since we can always explore further relations and implications for experience and because conditions change over time. How we come at this inquiry process also depends on our values, interests, and assumptions, which can themselves be subject to scrutiny and reconstruction in light of new conditions. In other words, our self as presently formed profoundly shapes our capacity to experience meaning and to know, and it directs our future inquiries.

Dewey often formulates this emergence of knowledge possibilities as the transformation of *cause/effect* relations into *means/consequences* relations. Causes and effects are not ontologically distinct kinds of events; they are merely phases of a temporally developing experience. We pick out some events and call them "causes" insofar as we think they influence what follows them, which we call "effects." But effects are events that themselves can then play a role (as causes) in future effects. So, any given event might be an effect of many prior causes, but that effect might be a cause of some subsequent event. "Causes" and "effects" are useful discriminations that we make within an ongoing flow of experience in order to help determine how we might alter the course of some developing experience. There is no single, unified cause of any particular event, since there are always multiple conditions of any event. We pick out one single event (E_1)—usually one that we think we can manipulate in order to influence future events—and we call it "*the* cause of E_2," as if it were single-handedly responsible for generating E_2.

Means and consequences are causes and effects viewed from the perspective of their meaning for our lives. The meaning of a given object or event is revealed through both its antecedent conditions and its possible consequences. Therefore, to grasp the meaning of some object or event is to see it as its own development of possibilities latent in prior conditions and as a means connected with certain consequences. Like causes and effects, means and consequences are aspects of one continuously developing situation.

An important implication of this idea that knowing is a remaking of the world is that scientific laws concerning individual objects or events are not universal statements of preexisting, fixed relations among things. Instead, "laws are intellectual instrumentalities by which that individual object is instituted and its meaning determined" (Dewey 1925/1981, 164). Laws are tools (technologies) or methods for conducting inquiry. They are what we might call guiding hypotheses, that is, "*formulae for the prediction of the probability of an observable occurrence*" (1929/1984, 165).

The beginning of knowledge, then, is an appreciation of means/consequences, that is, seeing the implications for experience (consequences) of present or past conditions (means). To know the meanings of a thing or event is to understand how that thing or event is likely to play out in experience—what its consequences are. To see certain events as possible means is to find a way to anticipate and affect the outcome of certain events (as consequences).

4.7   Knowing as Intelligent Experiential Transformation

Since knowing involves acting in and on the world, it is not merely a mental operation. It is a full-body operation that entails action in the world that transforms experience. It transcends mental versus physical, and mind versus world, because it never gets out of touch with the environment in the first place. Knowing does not just report on (i.e., represent) pre-existing conditions; rather, it is a form of action in which experience is transformed as our understanding of the meaning of our situation changes.

As we saw earlier, the chief error is dualism in traditional theories of knowledge that makes knowledge merely the re-presentation, in mind, of pre-existing, mind-independent states of affairs in the world. Dewey observes that such theories

> spring from the assumption that the true and valid object of knowledge is that which has being prior to and independent of the operations of knowing. They spring from the doctrine that knowledge is a grasp or beholding of reality without anything being done to modify its antecedent state—the doctrine which is the source of the separation of knowledge from practical activity. (Dewey 1929/1984, 157)

If knowledge is not intellectual reportage of this sort, then what is it? The answer is that knowing is a form of action in the world, by means of which an indeterminate, problematic situation comes to be transformed into one whose meaning is clarified and made determinate enough so that the organism can fulfill its needs and desires. In other words, we know something when our understanding of the meaning of that thing or event is relatively confirmed by our inquiry into its sources and consequences, thereby making it possible for us to function in our surroundings. In sum, knowledge "signifies events understood, events so discriminately penetrated by thought that mind is literally at home in them. . . . For it is directly concerned with not just instrumentalities, but instrumentalities at work effecting modifications of existence in behalf of conclusions that are reflectively preferred" (Dewey 1925/1981, 128).

When our knowing activities change the meaning of things, we then exist in a transfigured world. Dewey encapsulates this key idea that knowledge actually changes our world by imputing new meanings and characteristics to things: "Knowledge is not a distortion or perversion which confers upon its subject matter traits which do not belong to it, but is an act which confers upon non-cognitive material traits which did not previously belong to them, . . . characters, meanings and relations of meanings hitherto not possessed by them" (Dewey 1925/1981, 285).

It may be tempting to claim, as generations of philosophers have, that we can know our world because it manifests a fundamental rational structure. It should now be clear that Dewey rejects any such ontological and epistemic assumption because he saw that the terms "reason" and "rationality" have too often been hypostatized into pre-given, fixed rational structures supposedly existing in both mind and the world, thereby making it possible for mind to know, or correspond to, the rational structure of mind-independent nature. To avoid this tendency to reify rational processes into fixed structures, Dewey preferred the term "intelligence" as an activity of directed operations of inquiry utilized in our attempt to survive and flourish in our world. Intelligence is not a pre-given function or capacity; rather, it is an *achievement* in which the potentialities of a situation are realized in experience. He argues that "the worth of any object that lays claim to being an object of knowledge is dependent upon the intelligence employed in reaching it. In saying this, we must bear in mind that intelligence means operations actually performed in the modification of conditions, including all the guidance that is given by means of ideas, both direct and symbolic" (1929/1984, 160). *Knowing is intelligence at work transforming the world, not merely reporting on it.* At the same time, and by the same processes, it transforms the self that knows.

When Dewey says "intelligence in operation, another name for method, becomes the thing most worth winning" (1929/1984, 163), he emphasizes the transformative dimension of appropriate modes or methods of inquiry. "Knowing marks the conversion of undirected changes into changes directed toward an intended conclusion" (163). For example, our knowledge of home construction is actualized and enacted progressively as we lay the foundation, raise the walls, install the plumbing and wiring, put on the roof, hang the sheetrock, paint the walls, finish the woodwork, and more, leading up to the consummatory activity of someone safely inhabiting that house and making it a home. This kind of intelligent activity might be written off as mere "know-how," but even so-called theoretical knowledge has its inescapable practical components. For instance, you have knowledge of certain mathematical concepts only insofar as you can perform activities such as constructing proofs, tracing out entailments, and discovering new relations hitherto unknown. *All knowing is in some extended sense practical because it involves a change in experience.* And, as we move to higher levels of abstraction and thought, we might think of theoretical knowing as, in a way, a practical undertaking geared toward creating an existential "home" in the world, or perhaps as a therapeutic practice to heal various intellectual illnesses. At the level of theoretical reflections, the making, deepening, and enriching of meaning changes our relation to things and events.

All knowing is a doing. Sometimes it manifests in the manipulation of physical materials (brick, wood, tile, nails, wire, etc.) and sometimes in activities that use intellectual materials (i.e., meanings) that make possible acts of reflective inquiry, such as performing intellectual evaluations, studying the meaning of things and events, and making plans for the guidance of actions.

4.8   The Stages of Reflective Inquiry and the Reduction of Indeterminacy

The operations of knowing are temporally developing processes that move us, by stages, to renewed or recovered functionality and activity in the world. Not all knowledge is immediately and directly practical, but even our high-level theoretical inquiries are practical insofar as they shape the meaning of objects and events, thereby remaking some aspects of our world.

Roughly speaking, we identify four typical stages in the process by which we can move, via the need-search-satisfaction pattern, from discernment of a problem to its experienced resolution:

Experiencing a Problematic Situation

It is often overlooked that we do not engage in reflective thinking unless something goes wrong in our habit-shaped experience. The fact is that, for most of the mundane affairs of our ordinary lives, we can run on the autopilot of our acquired and deeply rooted habits of responding to our surroundings. No thinking is required, which is most of the time a good thing, since we would never get on with living if we had to think through every minute situation we encounter, or exercise conscious control over every bodily process.

Sometimes, however, we run into novel conditions, and then our habitual action is stymied. The prior expectancies we have acquired do not conform to what we experience. We are brought up short. We need to "stop and think" before we go on. Our current situation becomes indeterminate and confused. We feel or sense that our situation is problematic and we are not sure how to proceed. We need to recalibrate our expectations. Dewey's insight into the importance of unexpected events in leading to new knowledge construction proves useful as we consider the neural computational models of predictive coding (chapter 9).

Dewey argues that with the emergence of mind (and meaning and language), humans develop far more numerous and complex habits, and one fateful consequence of this is that a given situation may evoke multiple habitual responses, some of which are not compatible with others. The result is that we experience the meaning of an object or event as relatively indeterminate. We become aware of this indeterminacy as a feeling of lack of clarity and loss of confidence about what our situation (or some part of it) *means*.

Identifying a Problem

To begin with, we feel the indeterminacy of our situation, along with the anxiety generated by that indeterminacy, but we do not yet know it as such. We do not know what is wrong. The next step is to define the problem at hand, for this provides the basis for our directed activities of search for a solution. All the possible solutions we might imagine depend directly on what we take the problem to be, for this selective action determines what will be perceived as relevant to our problem solving. It is seldom obvious what the problem is, and we have to decide which characteristics of our situation are relevant and need to be attended to in our inquiries: "The risky character that pervades a situation as a whole is translated into an object of inquiry that locates what the trouble is, and hence facilitates projection of methods and means of dealing with it" (Dewey 1929/1984, 178). It is often assumed that the nature of the problem is evident and given, but Dewey observes that identifying the problem is itself a crucial step in the process of inquiry.

To illustrate, consider a mundane case where two people argue about how to manage a budget. One might think the problem is about rational actor economic considerations concerning how most efficiently to manage shared funds, while the other focuses on how they feel that they are not respected as an equal in the decision process. Which definition of the problem is right? Both are. The situation does not come with its preferred description or evaluation tattooed on its arm. Deciding what the problem is makes all the difference regarding which direction thought takes toward a solution. One might think that such indeterminacy exists in ordinary affairs of life, but not for theoretical problems of the natural sciences. However, it can be seen that the same indeterminacy applies in science and all our most rigorous modes of reflective thinking too.

Every scientific theory has to make some initial determination of what it is that needs to be explained, and it is widely recognized how much disagreement there can be about how to circumscribe the relevant phenomena. For example, say you study attention. Well, what are the relevant phenomena that need to be explained? For example, consider what is known as covert orienting (i.e., shifting attention to a new target, even though your eyes remain fixed on a present target). If you work with a "spotlight" model of attention, then covert orienting is a central phenomenon, akin to beginning to move a spotlight toward a new focal point, and it is assumed that there is an internal executive system to perform the attentional shift. However, if you are working within a causal resource model in which there is no homunculus or executive system that could initiate a refocusing of attention, then covert orienting doesn't play much of a role (Fernandez-Duque and Johnson 1999, 2002). There is no act of beginning to redirect one's attention, but only a resultant state of distribution of attentional resources. As we will see in chapter 6, the fundamental meta-phoric models you assume as you approach an inquiry will define what the relevant phenomena are and what terms are to be used to explain those phenomena. Consequently, even the identifying of relevant phenomena is itself a key part of the inquiry.

Or, think of the ongoing squabbles about which interpretations of quantum phenomena (such as quantum indeterminacy) are to be preferred. When considering the relevant light-diffraction phenomena to focus on, one chooses between the experimental results that support either the particle or the wave interpretation of light. There is nothing in the situation that defini-tively determines what the relevant phenomena are or how they are to be interpreted. In short, there is considerable relevance to the old saw "a problem well stated is half solved."

Imagining Solutions

The way the problem is defined, then, constrains the range of possible solu-tions available for consideration. Coming up with a range of possible solutions requires an ability to go beyond the inertia and entrenchment of sedimented habits of thought. In cognitive neuroscience, this is known as the stability/plasticity problem. Our prior experiences have reinforced prior habits so strongly that it is difficult to "rewire" our functional neural assemblies so that new possibilities can be considered. Dewey thought the best way to deal with this problem was to cultivate more habits that are complexly interwo-ven, so that new connections can emerge, rather than settle for a small set of fixed habits.

When facing indeterminacy, most of us look for quick, easy, simplistic solutions. We are not comfortable dwelling with the anxiety engendered by indeterminate situations and paucity of information. Anxiety does not feel good. We want a resolution of the indeterminacy, so that our uneasiness will be replaced with the felt release of tension that accompanies return to action. Dewey observes that we have a strong inclination to fasten on quick solutions that don't require sustained inquiry.

> The tendency to premature judgment, jumping at conclusions, excessive love of simplicity, making over of evidence to suit desire, taking the familiar for the clear, etc., all spring from confusing the feeling of certitude with a certified situation. . . . The natural man dislikes the dis-ease which accompanies the doubtful and is ready to take almost any means to end it. (Dewey 1929/1984, 181)

Anxiety and doubt make us uncomfortable. However, if we can tolerate it, that anxiety provides the impetus for intelligent reconstruction of our experience.

Feeling certain about something is not a mark of genuine intelligence. Feeling certain is merely a psychological state, whereas a certified situation is the product of intelligence actively exploring the meaning of that situation, testing out one's view in light of experience in order to resolve the initial indeterminacy encountered and to initiate a new understanding of one's situation. Dewey goes so far as to say that "the scientific attitude may almost be defined as that which is capable of enjoying the doubtful; scientific method is, in one aspect, a technique for making a productive use of doubt by converting it into operations of definite inquiry" (Dewey 1929/1984, 182).

Intelligent inquiry requires an imaginative projection of ways to resolve some of the indeterminacy of the situation in which we find ourselves to be confused and unclear. What this amounts to is exploring the *meaning* of our present situation—how it might be developed or transformed into a more determinate state in a way that enhances our ability to function well in our surroundings. We explore possible solutions, not in the sense of some final or complete resolution of the indeterminacy, but rather to reduce as far as we can the indeterminacy experienced *here and now*, realizing that every solution will at best be partial and often temporary, subject to changed conditions that require further inquiry.

Testing the Meaning of a Situation through Action

Finally, inquiry into the meaning of our situation leads to knowledge insofar as our new understanding allows us to function in our world. Through our inquiries, we generate new expectancies that we can test against forthcoming experience. Knowledge "marks a question answered, a difficulty disposed of, a confusion cleared up, an inconsistency reduced to coherence, a perplexity mastered. . . . Similarly, thinking is the actual transition from the problematic to the secure, as far as that is intentionally guided. . . . [K]nowledge is the completed resolution of the inherently indeterminate or doubtful" (Dewey 1929/1984, 181).

Knowledge is possible only as intelligence is active in the world. The source of the indeterminacy in a problematic situation is that any object or event is implicated in multiple highly complex webs of meaning, so that the meaning at any given time can be relatively indeterminate. Moreover, new circumstances can change the character of a situation. This indeterminacy can be resolved "only by actions which temporally reconstruct what is given and constitute a new object having both individuality and the internal coherence of continuity in a series" (Dewey 1929/1984, 189).

One's expectations constitute hypotheses about how experience will unfold under certain conditions. "Testing" a given hypothesis can never be an absolute, final achievement, valid henceforth and for all time, because conditions change and inquirers always have partial, often uncritical perspectives. Hence, the pragmatist insistence on the fallibility of any knowledge frame or conception. There can be "better" and "worse" solutions to a problem, just insofar as some hypotheses actually do a better job of reducing indeterminacy, harmonizing competing values, liberating our energies, and deepening our understanding, *but always for the moment at hand*. The intelligent strategy is ongoing critical inquiry, and our daily existence supplies more than enough opportunities for such reflective activity. Dewey quotes the biblical "Sufficient unto the day is the evil thereof," suggesting that we will never lack problems to address with all the resources for inquiry at our disposal.

This brief survey of the stages of knowing activity shows how the basic biological need-search-satisfaction pattern present in all living things continues to operate at the highest levels of intelligent knowing. Stage 1 is an instance of *need*—the felt sense of tension in the organism-environment transaction, due to the indeterminate character of the situation. Stage 2, framing the problem, is the initiation of *search* as effort to resolve the indeterminacy. Stage 3 is the continuation of that search by projecting expectancies and trying out in imagination various possible interpretations and courses of action. Stage 4 defines the *satisfaction* of the perceived need through action that transforms the character of experience and makes it possible for us to act more intelligently in the world.

4.9 Knowing as Embodied, Situated, Intelligent Action

To sum up, knowing is an inescapably embodied natural process for reconstructing experience: "The organs, instrumentalities and operations of knowing are inside nature, not outside. Hence, they are changes of what previously existed: the object of knowledge is a constructed, existentially produced object" (Dewey 1929/1984, 168). What is constructed is both a new situation and a new self, both emerging in that situation; not two separate things, but rather dimensions of a single complex interactive natural process. As mind discovers new relations, new consequences, new possibilities for action, it transforms the world as a scene of action. At the same time, the mind is itself transformed.

Dewey summarizes the import of embedding knowing in nature:

> When an interaction intervenes which directs the course of change, the scene of natural interaction has a new quality and dimension. This added type of interaction is intelligence. The intelligent activity of man is not something brought to bear upon nature from without; it is nature realizing its own potentialities in behalf of a fuller and richer issue of events. Intelligence within nature means liberation and expansion, as reason outside of nature means fixation and restriction. (Dewey 1929/1984, 171)

Because nature is in process, and because knowing is an activity of change and development of nature, there can be no finality of inquiry, nor any ultimate escape from the anxiety of doubt. The best we can hope for is an ability to use our doubt and anxiety to explore the meaning of our situation and try out new understandings. Because there is no fixed or ultimate reality, our best hope for living well and rightly must be to cultivate critical and imaginative intelligence. Dewey concludes that faith in intelligence is crucial for our ability to address the problems of everyday life:

> Because intelligence is critical method applied to goods of belief, appreciation and conduct, so as to construct freer and more secure goods, turning assent and assertion into free communication of shareable meanings, turning feeling into ordered and liberal sense, turning reaction into response, it is the reasonable object of our deepest faith and loyalty, the stay and support of all reasonable hopes. (Dewey 1925/1981, 325)

As Dewey articulated his pragmatist perspective on the mind at the opening of the twentieth century, the science of psychology was beginning to apply the scientific method to understand the mind. Although it was essential to the pragmatist approach to keep philosophical inquiry open to the advances in science, for the most part, philosophy and psychology went separate ways in the twentieth century. In the next chapter, we consider the challenges of a scientific study of the mind, and the schism that opened up in university culture between humanities and sciences, as these developments set the context for today's opportunities for a natural philosophy of mind.

5 The Challenge of a Meaningful Science of Mind: The Quest for an Objective Human Science

Science is a way of knowing. Originally, science *meant* the discipline of knowing. Now we define science not so much by the conscious process of knowing, as in classical philosophy, but rather by the scientific method through which we attempt to formulate clear ideas (*hypotheses*) and then test these ideas against objective evidence. Based on methods of rational analysis and empirical testing, the physical sciences have come to be regarded by many as the pinnacle of objective inquiry. Even better, when the evidence is made quantitative, we develop mathematical formulations that frame the regularities of the world in explicit detail. These mathematical models, such as Newton's laws of motion, can be applied to new situations with precision. This ideal of mathematical objectivity is the context within which scientific theories of knowledge were assumed to be constructed.

When psychology broke away from philosophy to become an experimental science in the nineteenth century, deep concerns arose about whether a science of the human mind could achieve as much objectivity as the physical sciences. For example, it was believed that physics could formulate universal laws governing the deterministic causal relations of physical particles. However, for those who assumed that human mental processes and behaviors are consequences of free will, it became unclear whether the social sciences could ever formulate the kind of causal laws governing social phenomena that supposedly gave precision and objectivity to the natural sciences.

A fundamental split emerged between what philosophers called the *Naturwissenshaften* (natural sciences) and the *Geisteswissenschaften* (sciences of mind/spirit—the human sciences). The former pursued deterministic causal laws governing all physical phenomena, while the latter claimed that a human science required a different method—a method of interpreting the meaning of human thoughts, values, and actions. This alleged difference between causal and interpretive explanations gave rise to what C. P. Snow (1959) called the two cultures of the academy: the natural sciences versus the humanities. The natural sciences came to pride themselves on their use of scientific method to predict events, while the humanists claimed that only they could adequately account for the *meaning* of those events for our lives.

For most of the first half of the twentieth century, there ensued a lively debate among philosophers of science between those who believed a unified science of everything was possible, and those who thought that the apparent uniqueness of human mind and freedom necessitated a different, noncausal, interpretive method of explanation.

The burgeoning science of psychology found itself in a no-man's-land between the objective methods of the natural sciences and the subjective (meaning-based) focus of the humanities. Seeking to demonstrate its scientific objectivity, psychology focused on eliminating subjective bias in its methods of inquiry. This took the form of replacing introspective methods with supposedly objective description and experimental testing based on observable conditions and behaviors (i.e., stimuli and responses). There was also a strong tendency to shy away from subjective aspects of cognition, such as emotions, feelings, and even conscious states. In short, to be good scientists, psychologists decided they had to avoid subjectivity in both the methods of scientific inquiry and the phenomena studied.

The result was the assumption that a university education could follow either objective science or subjective self-awareness—take your pick. This adoption of a rigidly objective mind set, with no room for subjective experience, was accepted uncritically by each new generation of students of psychology. Personal experience was deemed to be mostly irrelevant to the scientific study of human nature. This perspective has too often shaped cognitive science and cognitive neuroscience, and it has contributed to the alleged gap between objectivity in science and subjectivity in the humanities that still too often defines the two cultures of the academy.

For the educated person in the modern world, science has irrevocably shaped the process of knowing (Whitehead 1933). We can ask questions about the world and receive answers by evaluating scientific evidence with critical thinking. Yet, science is important not only to understand the outside world, but also to understand the workings of our minds. In fact, a meaning-ful science of mind may *require* self-consciousness, awareness of emotions and feelings, and attention to visceral values. A key principle of this book is that a scientific study of the mind must explain the subjective process that is integral to the activity of knowing.

A brief survey of psychology in the twentieth century will show how difficult it has been to achieve such a meaningful mind science that does justice to the subjective aspects of mind while also providing objective eval-uation of experimental evidence. In contrast, at the end of the nineteenth century, the American Pragmatist psychologists and philosophers, includ-ing William James and John Dewey, imagined such a science of conscious experience—a discipline of knowing that included the subjective per-spective on the mind. James's amazing *Principles of Psychology* (1890/1950) is a marvel of scientific research combined with some of the most cap-tivating phenomenological descriptions of our experience of perception, thought, and action ever penned. But the science of the mind that emerged in the twentieth century gave up consciousness—and the mind's subjective aspect—in order to aspire to the supposed pristine objectivity of the scien-tific method.

If the science of psychology did a poor job over the last century with understanding consciousness, it did a worse job of explaining the uncon-scious mind. Although modern scientific research provides abundant evi-dence that most of our mental activity is unconscious—in that we have little or no direct awareness of its mechanisms—this has been a fairly recent and even grudging recognition. When Sigmund Freud proposed that the majority of human experience emerges from a rich and dynamic uncon-scious mind, scientists were unified in rejecting his audacious proposal. This rejection occurred even as many humanists realized the paradox that understanding the basis of subjectivity meant gaining insight into the unconscious determinants of experience.

Psychology and early cognitive science adopted objectivity as a kind of pure ideal we are calling *pristine objectivity*. It is a continuation of the crav-ing for complete certainty that we considered in the philosopher's quest in chapter 2. To avoid subjective bias, psychologists thought they had to avoid emotional and motivational influences on cognition that are well known to impair objectivity. At the same time, these scientific approaches adopted the mistaken assumption that human thought occurs on a mental plane, completely different from the sensory and motor operations that are the biological foundation of the brain's functional systems (see Barsalou 1999 for a survey and critique of this view). The result was an overly objectified, disembodied cognitive science that has proven incapable of accounting for the depth and richness of human meaning, thought, and consciousness (Varela, Thompson, & Rosch 1991).

The disembodied approach to cognitive science aspired to an ideal of pristine objectivity, unfettered by personal needs and subjective awareness, which resembles Plato's idealism. If we are right that an accurate scientific understanding of the mind must accept its embodied biological and social nature, then most twentieth-century cognitive science remained with Plato in the cave, straining for a fleeting glimpse of the ultimate forms or ideas that supposedly make objective knowledge possible. We will argue that coming out of the cave means realizing that the mind can only emerge from the self—a self embodied in and engaged with nature. Escaping the cave does not mean coming to "see" the timeless essences of things, but rather dwelling and acting intelligently in the natural world—seeing ourselves as social animals enmeshed in our physical, interpersonal, and cultural surroundings.

The struggles of scientific self-awareness, and the understanding of our biological foundations, are striking when put in historical perspective. A brief overview of these struggles provides an important context as we consider a natural philosophy of mind—what we describe as a second-generation cognitive science—embodied in the biological mechanisms of an evolved brain. These new ways of thinking suggest opportunities for integrating science and humanities that may be particularly important to our current global society as we face the possibilities of transformations of the human condition by our rapidly unfolding information technologies.

5.1   Toward a More Meaningful Cognitive Science

In a meaningful science of knowing, we must know ourselves. The process of knowing involves *both* what is known and the acts of the knower. In our study of the embodied mind, we will see that your *self*—the net result of your personal history—is the ever-present agent in the process of knowing. *Knowing can then be understood as a process of self-organization.* The integral role of the self will be a central theme in the biological theory of mind, and in our psychological and philosophical interpretations, in the chapters ahead.

The problem is that applying science to ourselves is intrinsically more difficult than applying it to the properties of the external world. As we become self-aware, knowing becomes a *recursive* process, in which the experience of knowing is applied to understanding the experience of knowing. It is like pointing the video camera of your phone toward a mirror: the image can be lost in a recursive loop.

Now, logically, self-awareness seems like a natural advantage in a study of the mind, because each of us has some degree of subjective access to the mind's workings, and we should be able to put that to good use. As we come to understand our subjective biases, we should then be better able to put them aside and adopt a more objective attitude toward the process of knowing. The evidence on the mind's workings—from careful observations of people's behavior, structured laboratory experiments, and measures of brain mechanisms—could illuminate an objective understanding, both of people in general and of subjective experience of one's own mind. Indeed, bringing an objective analysis of the brain and cognition to clarify the process of knowing in personal experience is the central goal of this book.

Unfortunately, the mental discipline for maintaining both subjective and objective perspectives has proven elusive. Even though the study of the mind in philosophy started and ended with subjective experience, as psychologists adopted what they took to be scientific method, they decided that in order to be truly unbiased and objective, they had to give up considering subjective experience altogether and rely only on objective evidence. This attitude may change with the emergence of a suitable natural philosophy, but it remains stubbornly ingrained in the cognitive science and neuroscience still dominant today.

Undergraduate psychology students learn about pristine objectivity—implicitly of course—early on in their course work. They learn that psychology is about objective research. Questions about subjective experience are treated as embarrassing personal disclosures. What professors may not appreciate is that when psychology is only about research, and does not inform the subjective perspective, the process of knowing may remain naive, not fundamentally different from that of the layperson.

Important progress has been made in recent years to bring a more holistic understanding to cognitive neuroscience, computational science, and neurophilosophy. We can now include unconscious motives, mechanisms of sleep that integrate each day's experience with the personal history of the self, the role of emotions and feelings in cognition, and the subjective experience of consciousness itself. In what we call a second-generation cognitive science (Lakoff & Johnson 1999), these dimensions are recognized as essential elements of the emerging natural philosophy of mind.

It may be none too soon. In these early days of the third millennium, artificial intelligence (AI) promises to surpass human intelligence in important ways. Our machines may soon talk back. Our children and grandchildren will likely be presented with new modes of neurocybernetic fusion, digitally augmenting their physical and intellectual capacities. To appreciate what's happening, we may need subjective as well as objective insight into what it means to be cognizant. To remain subjectively naive may be increasingly dangerous in the Information Age. Those who make decisions will too often be informed only about the technical specifics of the powerful new capacities of machine intelligence because their intellectual work has not been grounded in humanistic principles and the rich appreciation of values that emerge from a well-rounded education.

To document the challenges of scientific self-awareness, we consider some problems that scientific psychologists have had in dealing with the subjective aspects of mind.

5.2   The Natural Subjective Perspective in American Pragmatism

It didn't start out this way, with objectivism running the show. For example, the American psychologist and philosopher William James looked forward to the twentieth century, assuming that many domains of the mind—meaning, consciousness, attention, language, religious belief—could be approached scientifically through rational analysis and laboratory experiments. At the same time, the new science of psychology was assumed to be meaningful, giving insight into their experience in personal terms. Even though there have been many findings in the century since James's *Principles of Psychology* (1890/1950), this openness to the natural questions concerning what it means to have a mind still makes James's writing a model for teaching psychology today.

James's appreciation of the everyday process of experience, his brilliant phenomenological descriptions of mental processes, and his conviction in the value of scientific analysis have many close parallels with John Dewey's argument that the philosophical study of human experience is fully compatible with scientific methods. Even as we encounter forms and methods of knowledge that were not available a century ago, we can hope to regain the open minds and enthusiasm that characterized the pragmatists' engagement with science and philosophy. In chapter 4, we presented classical pragmatism as offering a clear model of what a meaningful mind science would look like.

5.3   Freudian Insight, Hubris, and the Rejection of Psychoanalysis

The fragmentation of our concepts of mind in the early twentieth century is perhaps all the more remarkable because there were reasons to believe that a new synthesis of the scientific understanding of human nature was appearing at the end of the nineteenth century. The main reason for this emerging synthesis was Darwin's achievements in studying human nature on the basis of biological evidence.

Perhaps the most influential student of the new biological evidence was Sigmund Freud. Freud was a neurologist who began his work by theorizing about the functional connections among neurons in terms that are still relevant (Freud 1895; Pribram & Gill 1976; Tucker, Luu, & Pribram 1995). With the Darwinian explanation of human origins fresh in mind, Freud explained how human motivations could be derived from animal needs and urges and yet still be controlled to fit the expectations of society.

The novel starting point was the realization that human nature could be understood as derived from animal nature. In one of his first psychological studies, Freud looked to the mental content of dreams as manifesting basic emotional desires—what he called *wish fulfillment*. Freud (1953) speculated that perhaps the bizarre imagery of dreams could be interpreted to reflect the personal mind's unconscious processes.

Like many of his speculations, Freud's interpretation of dream symbolism has not received much support from research. For example, there is little evidence that we use symbols in dreams to disguise urges or feelings that would otherwise be unacceptable. Nonetheless, even if Freud's initial theories were wrong, he asked the right questions. Recent evidence from research on sleep and dreams continues to emphasize the importance of each night's dreams in integrating daily emotional experiences with significant personal memories of the past (Walker 2009). We still don't know exactly how this occurs, but the rapidly developing scientific literature on the neurophysiology of memory consolidation is starting to provide intriguing clues.

What was most significant was Freud's recognition that the mind is more than the conscious contents that we can reflect on easily and report verbally. The clear implication is that the mind is largely unconscious. This major insight continues to be affirmed by modern psychological and neuroscience studies. Furthermore, in interpreting the unconscious foundations of human behavior—dreams, slips of the tongue, and symptoms of mental illness—Freud provided the first scientific study of how some of the motives that direct our actions emerge from the unconscious mind. These fundamental motive controls on the mind's process—the regulation of the mind by its base in emotion and motivation—may not be accessible to self-awareness without explicit training.

Psychoanalysis provided a method for this training. It is a method for psychotherapy, but it is also a method for training the analyst and, at the same time, researching the principles of psychology. To become a serious student of the mind, Freud taught that you must first understand how your mind has been formed through your personal history.

The discovery of the unconscious mind in psychoanalysis is what the philosopher of information Luciano Floridi (2014) describes as the third revolution in human understanding. This is the realization that we are not in control of our minds in the simple and direct way that we might have assumed. Like the Copernican and Darwinian revolutions before it, this third (Freudian) revolution brought insight about the nature of the world and humility about our limited place in it. Copernicus showed us that we are not the center of the universe; Darwin showed us our kinship with what we have regarded as "lower" animals. Freud then informed us that we are not the grand rational beings we have assumed we were, driven instead mostly by unconscious drives over which we have little conscious control. Floridi points out how influential Freud's discovery of the unconscious mind was on both the science and literature of the twentieth century. These revolutions dealt three decisive blows to our exalted sense of ourselves as free, autonomous, rational creatures in control of our world and holding a privileged place by virtue of our rationality.

Ironically, in spite of widespread dismissal of Freud's theories by university scientists, Freud's emphasis on the importance of the unconscious mind remains consistent with scientific evidence on the unconscious foundations of human cognition. As much as today's cognitive scientists would reject Freud's theories as overly speculative and unsupported by evidence, the evidence demonstrates clearly that our direct conscious insight into our own cognitive processes is highly limited, and that the essential operations of attention, concept formation, and memory consolidation occur in the mind's unconscious background (Damasio 1999, 2010; Dehaene, Changeux, Naccache, Sackur, & Sergent 2006; Posner & Rothbart 2000). To understand what it means to be conscious and rational, we need to appreciate the mind's basis in its unconscious mechanisms of motive control.

There were many scientific research studies in the first half of the twentieth century that attempted to test predictions of psychoanalytic theory. A number of physicians and psychologists worked to make psychoanalysis an experimental as well as observational science. However, as it became increasingly popular in the public press, Freud's psychoanalytic theory suffered from the hubris brought on by its own success. It became a kind of movement, creating a self-righteous dogma, rather than a scientific theory that was open to revision and change in light of new evidence or theoretical insight.

Freud's theory was championed by humanities scholars interested in interpretive methods for understanding human experience and values. They saw his theory as standing in direct opposition to what they regarded as reductivist scientific theories.

In the midst of these controversies, many people also didn't like the idea of psychoanalysts pointing out that their behavior could be explained by childhood emotional traumas and deprivations that they would rather forget. Reactions against the smug proclamations of psychoanalysis were in fact highly influential in articulating the reactive posture of a logic of pristine objectivity in science.

5.4   Unconscious Bias and the Critical Logic of Objective Truth

Karl Popper became an influential philosopher of science by proposing a logic that he believed made it possible for science to achieve objectivity. He reports that he was goaded into his search for an acceptable model of scientific inquiry by his irritation in dealing with a psychoanalyst (Popper 1959). The psychoanalysts of the early twentieth century became skilled at deflecting criticisms by interpreting the criticisms as personal biases, "defense mechanisms," of the critic. Popper recognized that it was unscientific to assert that psychoanalytic theory is true because it was accepted by experts, even an expert as august as Sigmund Freud. He argued that psychoanalysts' confident interpretations exemplified *confirmation bias*, cherry-picking evidence to confirm assumptions.

Popper argued that no empirical theory can ever be absolutely confirmed through induction based on existing evidence because it always remains possible that new conditions might arise that are unanticipated by, and incompatible with, the proposed theoretical explanation. If you provide "confirming" evidence, this might support the theory in a limited way, but it is not definitive proof. There could be other conditions in which the hypothesis or theory is *not* true, so that even with confirming evidence, there is no guarantee that the theory is universally true (Hempel 1965; Popper 1959). Instead, Popper (1959) championed the importance of *falsification* as the core method of science. If you provide evidence that a theory is wrong—that it cannot account for a particular case—then, Popper argued, this was the more valuable contribution to science because it constitutes *definitive* evidence that the theory is *not* universally true, and this serves as a spur to further inquiry.

Many scientists continue to embrace Popper's critical attitude, and they relish the chance to test a theory because they consider the continuing ability of a theory to withstand falsification to be the best evidence for the merit of the theory. However, as Hempel (1965), Kuhn (1962), Quine (1951), and others argued, it turns out that even in the face of anomalous evidence, a criticized theory can often be saved by revising one's assumptions about the status of evidence or which values of inquiry are taken to be most important. Quine even argued that, in extreme cases, one might even be willing to revise one's logic in order to save an explanatory hypothesis. Critical tests are essential to progress, as new hypotheses are entertained by a few with hope, by many with skepticism, and then put to experimental test. In Anglo-American academic psychology, the negative reaction to psychoanalysis and other theories of the mind resulted in the extreme skepticism of behaviorism.

Skepticism is certainly a healthy part of science, but in the discipline of psychology, it resulted in a complete rejection of the subjective perspective. This rejection ran through cognitive psychology, cognitive science, and, until recently, cognitive neuroscience. In contrast, subjective experience was a core concern of psychoanalytic theory. Freud reasoned that the mind must emerge from basic biological needs and drives. When these needs and drives operate on the mind's process directly, we think of pleasurable things. Freud called this *primary process* cognition. It is exemplified by fantasy, in which urges directly shape the mind's process. *Secondary process* cognition is thinking that is constrained by the demands of society, such that our needs and urges must be negotiated in relation to societal values and practices.

Freud's dynamic view of the mind resonated with the intuitions about human nature held by many humanists. As a result, Freudian theory became central in the humanities camp of the two cultures of the academy (N. O. Brown 1959; Marcuse 1941), even as it was rejected by those in the science camp.

Without doubt, there were egregious errors in psychoanalytic theory. When Freud described to the medical community of Vienna that his female patients reported being sexually abused by the men in their family, the outrage among distinguished physicians threatened to drive him from the medical community (Masson 2003). To rationalize what his patients reported, Freud—always the creative interpreter—speculated that perhaps the women only *fantasized* sexual contact with their fathers, stepfathers, and uncles. Freud thereby established the Oedipal complex, almost certainly the most stupid and misleading psychological theory of modern times. It was nonetheless endorsed not only by many academics, but also by generations of therapists and their patients. We need only to recognize the statistical frequency of today's sexual abuse of children to understand that the dirty secrets of male-dominated Viennese society were not fantasy but reality. As a result, Freud not only damaged what he began with psychoanalysis but caused therapists to mislead generations of trusting patients.

In spite of the serious mistakes and shortcomings, we would do well to appreciate the central goal of psychoanalytic theory and psychotherapy: to increase conscious self-awareness of the workings of the unconscious mind that organizes each instance of experience and behavior.

5.5   Psychology's Ideal of Pristine Objectivity

As psychologists struggled to assume the trappings of "real" science, they faced the problem that it is difficult to study the mind in the laboratory. This is true even for its component capacities, such as attention, memory, or perception. Moreover, it seemed as if these processes were not accessible by observation, and they were therefore deemed inappropriate topics for rigorous, objective empirical science.

The result was a movement emphasizing studies of animal behavior that came to be known as *behaviorism*. Creative animal behaviorists such as John Watson took their cue from animal trainers, such as those working in vaudeville and the circus. These trainers demonstrated impressive control over an animal's behavior with training methods that involved providing rewards and punishments to the animals in a consistent and systematic fashion. Watson claimed the same principles could be used to train children.

Studies of animal learning became popular research topics for academic psychologists. This research created an important body of knowledge about the mechanisms of learning that then was later informed and expanded by studies of the corresponding brain mechanisms. The success in manipulating overt behavior, including studies in humans as well as animals, was achieved at the expense of any inferences about the internal workings of the mind that might be relevant to that behavior. In fact, many behaviorists proclaimed that those internal mental processes were not only outside the reach of scientific experiments but, worse, not relevant to explaining behavior in scientific terms. We were left with "black box" psychology— there were observable sensory inputs and behavioral outputs, but nothing could be known about the internal workings of mind, which were therefore regarded as irrelevant for a science of mind. The pristine objectivity of behaviorist methods rested on their exclusive reliance on observable, measurable behaviors and environmental conditions.

Even as the humanists were working away in their camp, exploring the implications of psychoanalytic theory for sexuality, art, politics, and the basic questions of human nature in mid-century Western culture, the behaviorist psychologists were equally busy in their camp, asserting that the mind as traditionally considered was an unnecessary illusion. The process of knowing was a real experience for humanists, but it seemed to evaporate as a topic of science.

5.6   Machines Like Us

By midcentury, a problem arising from an unexpected source challenged the success of behaviorism. Building on engineering principles of feedback control, digital memory storage, and logical instruction sets, electronics engineers made progress in building useful computing machines. The practicality of electronic data processing (EDP) soon became undeniable in many fields. These included not only obvious applications to business accounting and finance, but also several novel applications such as missile control and airline reservations.

For academic psychologists, the problem that arose was that the new EDP machines were regularly demonstrating capacities—such as calculations, memory, and sequential ordering of actions—that they had confidently determined could not be studied in the human mind! In response, some psychological researchers began studying processes such as perception or decision making under the rubric of *human information processing*. The specificity of formulation in computational instructions seemed adequate assurance of rigor and objectivity in considering the hidden, cognitive mechanisms of the human mind. We implicitly recognized the emergence of intelligence in the machines that we had—unconsciously of course—created in our own image.

5.7 First-Generation Cognitive Science

As psychology departments formed at universities in the early decades of the twentieth century, many founders split off from philosophy departments and their humanistic methods. In the later decades, a similar exodus led the more rigorous scientists (at least by their own definition) to leave psychology to form departments of *cognitive science*. In a positive sense, the cognitive scientists adopted the laboratory rigor of experimental psychology and the powerfully analytic methods of both computer science and linguistics. It was these structured laboratory methods and analytic descriptive frameworks that, with the emergence of neuroimaging, allowed the development of today's *cognitive neuroscience*.

For our purposes, however, it is also important to recognize the negative attitudes and the assumption of pristine objectivity that were characteristic of this early conception of cognitive science. By creating this field, scientists rejected the "soft" disciplines of psychology, including research into emotion, social interaction, and personality. These approaches to psychology were not allowed in the new departments of cognitive science because they bordered dangerously close to the humanist ghettos of the academy.

The chasm of the two cultures of the academy continued to split university intellectuals, as the "softer" parts of psychology and other social sciences were relegated to the lowly arts and humanities, no longer allowed among the allegedly rigorous and objective sciences.

The integration of mainstream generative linguistics (Chomsky) within cognitive science was a natural step because the view of language as a formal system could be adapted to the rational, objectivist program that cognitive science adopted. The Chomskyan conception of language as based on an innate rule-governed set of formal operations seemed to fit well with the emerging computational theories of cognitive processing.

The idea that the mind's operations could be specified in the same way as a computer program became a highly attractive component of the cognitive science agenda. This "disembodied" approach to mind, thought, and language has been labeled *first-generation cognitive science* (Lakoff & Johnson 1999).

The increasing importance of computer technologies during the late twentieth century led many cognitive scientists to believe that studying the mind as an analytic computational engine would prove the most effective paradigm. They accept as a literal truth the metaphor THE MIND IS A COMPUTER PROGRAM. The human information processing tradition continued through the formation of departments of cognitive science. There was widespread belief that AI would rapidly develop through this approach, to the extent that the mind's operations could be modeled via the digital logic of computational systems.

However, the continuing failure of AI as implemented with symbolic logic, such as the LISP language, frustrated this approach. Instead, the advances in AI in the twenty-first century have been with artificial neural network (ANN) models, in which patterns and knowledge are represented not with logical symbol systems (as in LISP), but rather with the distributed representations of brain-like networks. As we examine the insights from ANN models in a later section, it will become apparent that advances in AI required a different way of thinking about intelligence than the highly rational logic of traditional symbolic AI.

Analytic logic has indeed been productive for the sophisticated software of modern digital computers. But that logic proved to be insufficient. Cognitive science assumed that the mind's essential nature could be captured with digital logic—as if the logical structure of language were the essential backbone of the mind's operations. Instead, the theoretical success of distributed ANN (connectionist) models of the 1980s (Rumelhart & McClelland 1986), and the practical advances in AI with the implementation of ANNs in the early 2000s, provided a different, and important, lesson about the mind. Even though language is a powerful capacity of the human intellect, the mind's core mechanisms are not primarily linguistic. These emerge from the distributed neural connections (the neural networks), with patterns of intelligence that have evolved through the progression of the vertebrate neural architecture. Language is built by recruitment of these deeper embodied cognitive structures and processes of meaning and thought. Therefore, language is derivative, rather than foundational, for human cognition.

The failure of symbolic models of AI is significant for the failure of first-generation cognitive science generally. More than that, it calls into question whether science can illuminate human cognition and behavior if it operates under the assumption of pristine objectivity.

5.8 Second-Generation Cognitive Science: From Cognitive Neuroscience to Natural Philosophy

The last several decades have seen major advances in scientific understanding of the human brain's venerable neural architecture. To be fair, research into the brain mechanisms of behavior progressed during the behaviorist era of the twentieth century. Measures of the brain's electrical activity, such as with electrodes implanted in the rat's brain, provided objective, scientific data. Brain systems were interpreted largely with behaviorist concepts, including sensory and motor control of overt behavior, but also mechanisms of reinforcement that were thought to explain learning. With the "cognitive revolution" of human information-processing research midcentury, it became fashionable to investigate neural mechanisms of information processing, with concepts drawn from the traditional faculties of the mind, including perception, memory, and attention.

Throughout the twentieth century, sophisticated theories of brain function developed from a long tradition of clinical observations of the effects of brain damage on human mental function. However, the founding of the new field of cognitive neuroscience in the 1980s came about primarily through the development of new methods of *neuroimaging*, imaging brain activity with positron emission tomography, and soon functional magnetic resonance imaging (fMRI).

All of a sudden, it began to seem possible to explore the inner operations of mind that are mostly inaccessible to consciousness. This led philosopher Patricia Churchland to envision what she called the "co-evolution" of philosophy and cognitive science, working together in search of a suitably nonreductive and scientifically responsible theory of mind (1986, 373ff.). At the same time, the field of embodied cognition theory (to be discussed in chapter 6) began to challenge the first-generation disembodied cognitive science framework with empirical research from linguistics, developmental and cognitive psychology, phenomenology, and neuroscience. The result was a *second-generation "embodied" cognitive science*, grounded on the emerging research on how our brains and bodies give rise to meaning, thought, values, emotions, and reason (Feldman 2006; Lakoff & Johnson 1999).

Particularly in the early years, the technical challenges of measuring brain function with neuroimaging technologies became scientists' primary concern. Yet, certain careful observers, such as Marc Raichle at Washington University at St. Louis, realized that imaging the brain's activity could be related to psychological processes only if you had a careful specification of *what* psychological process occurred during the imaging process. Raichle recruited Michael Posner from the University of Oregon to help with this specification of cognition. Posner was well known for studying cognition with the *chronometric* method, which involved measuring the time required for a specific cognitive operation, such as shifting attention from one region of the visual field to another. The measuring of time became a way of capturing the specific mental process, defining a physical property of cognition.

An even more analytic method was to break down what are naturally complex and multidimensional cognitive processes into simpler ones using the *subtractive method*. A complex task, such as paying attention to a visual target, could be broken down into specified components. The task of responding quickly to a target at an unexpected location could be contrasted with the same task, but now at an expected (attended) location in space. The faster reaction time for responding to the target at the attended location could then be attributed to the specific cognitive operation of consciously directing visual attention in space.

From these foundations in integrating laboratory methods of cognitive psychology with the advanced measures of brain activity from imaging brain metabolism and blood flow, cognitive neuroscience has developed into a complex field of its own. The two keys were advances in neuroimaging technology and clever experimental methods, mostly derived from cognitive psychology. Without doubt, important new concepts of the brain and mind's operation came from this work.

Paralleling the theoretical advances in understanding computational neural networks that developed into today's AI, the evidence collected by cognitive neuroscientists led to recognition of the *distributed* nature of the brain's cognitive functions. The human brain is now understood to be organized in *functional networks*, widely separated cortical networks that interact with each other on a regular basis, providing general systems of sensory and motor integration, memory organization, and abstract conceptualization. These and other crucial systems are described in chapters 7–9.

Perhaps the clearest evidence of an emerging natural philosophy of mind that is compatible with today's scientific research is the progress in understanding the neural mechanisms of consciousness. A topic such as consciousness seems to invite loose thinking—exactly what scientists have tried to avoid. Yet, a number of careful neuroscientists now bring both theoretical analysis and advanced empirical methodologies to understand how consciousness is supported by specific neurophysiological mechanisms (Damasio 1999, 2010).

For example, building on the insights from distributed neural networks, Giulio Tononi suggests that consciousness could be defined as the integration of information across the brain's representational systems (Tononi 2008; Tononi & Edelman 1998). Although expressed in the terms of information theory, this approach captures the idea of consciousness as the process of awareness that integrates the mind's current operations. Such a process would need support from the brain's working memory systems, so that it becomes a reasonable topic for scientific research to study how and when consciousness is supported by specific neurophysiological processes (Dehaene 2014).

As we gain more understanding of the physical mechanisms of the brain that are required for consciousness, there is also progress in simulating neural networks with machine intelligence. It is now reasonable to speculate whether sufficiently advanced ANNs could bring an equivalent form of consciousness to machines (Dehaene, Lau, & Kouider 2017).

5.9 Cognitive Neuroscience in Dialogue with the Humanities

Although the focus of this chapter is on the challenges of bringing humanistic questions to a modern science of the mind, it is worth recognizing that the increasing theoretical and philosophical breadth of modern neuroscience, exemplified in the writings of Damasio and Tononi, cited above, has captured the attention of creative humanities scholars. For example, Donald Wehrs and Thomas Blake (2017) have reviewed the modern cognitive neuroscience literature on the neural mechanisms of emotion and motivation in their recent study of the historical trends in literary criticism. The understanding of meaning conveyed in literature can be augmented by appreciating the mechanisms of human value and emotional response that have been important topics of theoretical advances in cognitive neuroscience. In the Wehrs and Blake volume, the historical account of what has been taken as significant in literature is enriched and illuminated by our present contrast of embodied cognition with the more rationalized model of scientific and philosophical thought we have described as pristine objectivity.

There are other encouraging examples of productive collaborations between neuroscience and the humanities. In *More than Cool Reason* (1989), George Lakoff and Mark Turner used second-generation cognitive science to analyze poetic meaning. In *Reading Minds: The Study of English in the Age of Cognitive Science* (1991), Turner expanded on this project, which later led to his collaboration with Gilles Fauconnier on *The Way We Think: Conceptual Blending and the Mind's Hidden Complexities* (Fauconnier and Turner 2002), which sought a comprehensive theory of how conceptual blending operates in many types of cognitive processes. Another exciting development is Edward Slingerland's *What Science Offers the Humanities: Integrating Body and Culture* (2008), which chronicles some of the major shortcomings of previous humanistic scholarship, and then explores how the humanities can be enriched through collaboration with cognitive science.

These examples illustrate that modern neuropsychological theory is beginning to be relevant to more than restricted laboratory tasks, and is now able to speak to more fundamental questions of human experience examined in humanities studies. Perhaps the chasm of the two cultures is no longer dividing disciplines as effectively as it did in the past. We may anticipate experiments in weaving together science and the humanities that are highly relevant to the prospects for a natural philosophy.

5.10   Artificial Intelligence and the Fifth Revolution

It may be particularly challenging to integrate the remarkable advances in AI with the humanistic perspective we are proposing is essential for a natural philosophy. Luciano Floridi's analysis has suggested that the transformation of the human condition with the advent of modern information technology amounts to a fourth revolution. He suggests that this fourth scientific revolution is as significant in redefining the place of humans in the cosmos as the Copernican, Darwinian, and Freudian revolutions.

Clearly we are seeing profound changes in society as a result of information technology, so the idea of a fourth revolution makes sense. Yet, at least so far, the science of AI has brought new engineering capabilities, but not much insight into who we are and how human intelligence actually works. The convergence of the AI of neural networks with human neuroscience may usher in a fifth revolution, as science illuminates the workings of subjective experience.The assumption that human intelligence can be understood as an objective and rational process led first-generation cognitive scientists to concern themselves with the logic of computer programming, as well as with experimental psychology and Chomskyan linguistics. As noted above, there was continuing interest in AI, in which symbolic languages such as LISP were used to simulate human cognition as a logical symbolic operation. Yet, the progress in AI with symbolic languages was poor. The success of more recent AI through neural simulations has emphasized that intelligence may not be limited to thought expressed in some allegedly objective logic of rational linguistic constructions—even an artificial computational language—but may be emergent in the biological pattern recognition that has evolved in the neural networks of brains.

We will trace the development of neural network AI carefully in chapters 7 and 8. For now, it is enough to note that as cognitive neuroscience developed rapidly in the late 1980s and early 1990s, there was a time when it seemed as if the discoveries about ANNs would provide the basis for new theoretical insights into the human brain, as these were applied to interpret the evidence from neuroimaging with fMRI and related technologies (Rumelhart & McClelland 1986). The ANNs provided appropriate metaphors for understanding the brain. Rather than thinking of the brain as a digital computer, with logical operations sequencing and transforming words in memory, the ANN achieves intuitive representations of the world (the input) through highly parallel neuronal networks.

For some reason, this theoretical work quickly faltered. Perhaps it was the focus on methodological specialization required for young scientists who took on neuroimaging experiments. Or perhaps it was the difficulty that the ANN modelers had in extracting general principles from their simulation studies. Whatever the reason, the fields split into a somewhat narrow empirical cognitive neuroscience and a mathematically focused engineering of computational simulation. The young scientists in each camp went to different conferences and published in different journals. As a result, the AI transforming our daily lives now is an engineering skill, developing rapidly by technologists with strong economic incentives but providing minimal scientific insight into our own cognitive processes.

It may be, then, that the coming advance in AI will progress on its own, with great impact on our lives to be sure, but with little contribution to understanding our own neural capacities. On the other hand, we might hold out hope that we will once again see that we have created machines in our own image. This time it is not the general computational model of information processing, but rather the specific architectures of distributed neural networks that have allowed advances in machine intelligence.

If so, then it may be recognizing the principles of distributed computing in our own cognitive operations that provides the insight required for a fifth revolution when we understand the process of mind in cybernetic terms. Until this happens, progress with AI remains a technology trick—something that transforms our world at the hands of technical engineers who have little training in humanistic principles and little insight into the subjective process of human minds. This may result from a century of striving for pristine objectivity in psychology and neuroscience, as the science of intelligence has been kept as a disembodied artifact, rather than a way to understand the conscious, subjective process of experience.

5.11   Realizing the Embodied Self

In machines and humans, if we are to spawn a fifth revolution, the functional understanding of mind must emphasize the integration of information for the whole system, whether this is a human brain or its AI simulation. For humans, conscious cognition is not an isolated action, separate from personality; rather, it is shaped by the mind in its entirety, including its developmental history that formed the cumulative memory of the self.

Without taking the whole person into account, we might assume that cognition in one domain, such as deciding whether to believe in global warming, may be limited to the evidence in that domain, such as the scientific data on warming trends. However, because each act of cognition is embodied within the whole personality, we need to realize that for some people, the message from scientists on global warming is taken as a threat to their worldview, challenging the conservative personal belief system broadly. For many conservative traditionalists, the self seems embedded in the familiar past of an unchanging world. How can we understand this incompatibility of objective knowledge with the core beliefs of the self?

When cognition is framed as discrete acts of information processing, as in first-generation cognitive science, we limit our consideration of human intelligence to the specific task or experimental manipulation under examination. Perhaps unexpectedly, when we reframed the information apparatus in the mechanisms of distributed neural networks, the necessity of a holistic view of mental representation came to the fore. This constitutes a profoundly important, almost revolutionary, change in our understanding of mind and knowing that begs for a fifth revolution. In many ways, we see that an appreciation of the mind in the context of the whole organism brings scientific clarity to the challenge of understanding the process of knowing as an operation of the self, such that gaining new insight is invariably self-transformational.

As we look further, in the remaining chapters, into the biological theory implied by the study of the mind's neural basis, we discover that the brain's networks are continually reorganized to maintain the ongoing continuity of the changing self. The subjective integrity of mind—in both its conscious and unconscious aspects—becomes an essential tenet of the science of the embodied mind, where embodiment is not just the situation of mind in bodily form, but also the holistic operation of a continually developing personality.

In this chapter we considered how the quest for absolute certainty of knowledge has been maintained over the last century, not only in the highly linguistic reasoning of analytic philosophy, but also in the rationalized objectivity of the linguistic, computational, and narrowly experimental research of first-generation cognitive science. The science of the mind has strived for rationality by assuming that the mind's process can be captured through semantic descriptions of it. In his extensive historical review of the intellectual progress of modern science, Alfred North Whitehead (1925/1997) pointed to this confusion of objective semantic description with the real meaning of things as the *fallacy of misplaced concreteness*. This notion is not unlike John Dewey's account of the *philosopher's fallacy* (1925/1981, ch. 1), of taking the ideas of the philosophy, which are always selective and value-based, as a sufficient account of the reality of self and world.

Because so many errors stem from assuming that human cognition is rooted in the structure and semantics of language, it might seem strange that a rich appreciation of the bodily basis of meaning and thought has come from the study of natural languages. Nonetheless, as early as the late 1970s a perspective known as cognitive linguistics began to supply extensive evidence that human conceptualization and reasoning are based on the patterns and processes of our bodily engagement with our world. In this approach, meaning and thought are not merely linguistic, semantic constructions. Instead, concepts are fundamentally rooted in our sensory, motor, and affective processes of experience, and language is built up from this embodied meaning. In this way, the embodied cognition theory that we examine in the next chapter reveals the bodily basis of meaning and how it is recruited for abstract conceptualization and reasoning, especially via conceptual metaphor.

6 Embodied Meaning and Thought

Before we delve into the neuroscience in subsequent chapters, we will consider evidence, primarily from the study of human language, that all conceptualization and reasoning are grounded in our bodies as they engage their environment. We explore evidence from cognitive linguistics (a subfield of embodied cognition theory), an approach that coalesced in the 1970s as a science-based alternative to the dominant Chomsky paradigm of language acquisition and competence. Against Chomsky's positing of an innate language (especially syntax) module that provides a universal basis for natural languages, cognitive linguistics presents evidence that our conceptualization, reasoning, and linguistic acts are based on preexisting capacities for perception, bodily movement, object manipulation, and emotional response, which are recruited for higher-level cognitive operations.

Thus, even though philosophers and scientists have often assumed that rationality is not shaped by our embodiment, the study of language itself reveals that human cognition is *embodied*, such that the very nature of our conceptualization, reasoning, and linguistic performance depends critically on how our bodies and brains are structured and how they interact with our physical and social surroundings. If our species had evolved substantially different bodies and brains, we would inhabit a substantially different world, and the ways we experience, understand, reason, and talk about that world would be quite different.

6.1 The Body in the Mind

In *Philosophy in the Flesh: The Embodied Mind and Its Challenge to Western Thought* (1999), George Lakoff and Mark Johnson presented evidence for three central claims of their embodied cognition version of cognitive linguistics:

(1) the mind is inherently embodied, (2) thought is mostly unconscious, and (3) abstract concepts are largely metaphorical. In this sense metaphors are not mere figures of speech, but rather the concepts arising from experience that give meaning to abstractions. Most of our meaning making and thinking goes on beneath our conscious awareness and is rooted in our sensory and motor capacities. This embodied meaning is profoundly shaped by our brain architecture and by the patterns and processes of our bodily engagement with our surroundings. These body-based, sensory-motor meaning resources are appropriated for abstract conceptualization and reasoning, mostly via imaginative processes such as conceptual metaphor (explained below). In other words, *all* of our meaning and understanding is embodied, either directly in patterns of physical interaction with our world or more indirectly through the recruitment of sensory, motor, and affective processes for abstract thought.

Lawrence Barsalou (1999) shows how widespread and deeply rooted the disembodied view of concepts really is. He analyzes and critiques the widely held commonsense theory of mind, according to which our experiences of physical objects, events, and actions are based on our sensory and motor capacities, but our *concepts* of those experiences are processed in entirely different parts of the brain that are not involved in actual perception and bodily movement. According to this view, given that concrete perceptual concepts are not grounded in the structures of our sensory-motor experience, then most assuredly our *abstract* concepts have nothing to do with the areas of our brain that construct perceptions and execute bodily actions.

Against this view that concepts are processed in regions of the brain entirely different from those that generate perception and action, embodied cognition theory denies any radical distinction between percepts and concepts. It provides evidence that *all* our concepts, both perceptual (concrete) and abstract, are grounded in the same brain structures and processes responsible for perception, bodily movements, and manipulations of objects in the world. Perceptual concepts are simply activations of neuronal clusters responsible for carrying off specific events of perception and bodily movements. Abstract concepts recruit these same bodily processes for higher-level understanding and reasoning.

It is in this strong sense that all our meaning is rooted in and grows from our bodily transactions with our world. Consequently, our meaning, thinking, knowing, and communicative acts depend on the same motivational controls and organism values that operate in the basic physical functioning of our bodies. Conceptualization and reasoning can never float free of, or transcend, our embodiment, though they can extend our embodied understanding in hitherto unexperienced situations in an imaginative expansion of meaning.

6.2   Image Schemas as Embodied Meaning Structures

To get a basic sense of what embodied meaning is and how it operates, let us begin with some of our most mundane daily experiences and actions in our surroundings. Before we have abstract concepts, before we develop propositional thinking, before we know how to reason, and before we acquire language, we dwell meaningfully in nature as physical organisms in ongoing contact with our environment. The nature of our perceptual systems, our capacities for action, and our bodily part-whole relations shape how we can interact with and make sense of our environment. Even as we mature to adulthood, we do not outgrow this bodily significance, but instead build upon it over the course of our lives. In fact, our higher-level cognition recruits these more primitive and basic structures and processes of embodied meaning making.

Consider, for example, the center-periphery structure of your perceptual world. From the perspective of your perceptual experience, your body is the center of your world, while your "world" constitutes a horizon of possible experiences, both past, present, and future. Merleau-Ponty (1962) described how, as we focus our attention on one object, spatial location, or event, the meaning of that object or event depends on an expansive context of connections and relations to other experiences that, although not currently attended to, constitute the horizon of potential resources for experience and meaning. John Dewey (1930/1981) called this enveloping background the *situation*, within which attended objects stand out and become meaningful. William James (1890/1950) observed that the meaning of a specific focal object depends on the surround of connections and relations he named the *fringe* of conscious attentional focus. This CENTER-PERIPHERY structure is thus immanently meaningful to us as a generic structure of all perception.

Other body-based spatial relations schemas arise from projections of body parts onto objects, such as the *foot* and *head* of a bed, the *foot* of a mountain, the *face* of a rock wall, and the *mouth* and *arm* of a river. The right-left symmetry of the body inclines you to project right-left orientation onto other objects and spaces (as in *right and left* field, the *right* and *left* sides of a book, and the *right* and *left* sides of the aisle).

A similar important body-based schema is FRONT-BACK orientation. We anthropomorphically project fronts and backs onto objects that have no intrinsic fronts or backs. Moving objects, such as buses, get a front projected onto them relative to the typical direction of their forward motion. Most of the time the bus moves *forward*, but sometimes it *backs* up. We tend to think of front-back orientation as intrinsic to objects when in fact it is defined relative to our bodies. For example, although a plain water bottle may have no intrinsic front, it can acquire a front as that surface that "faces" an observer, and a back as the unseen part on the opposite side of the bottle. However, if you turn the bottle horizontally and move it through space, it acquires a different front and back based on the direction of its motion. Likewise, if you roll a ball along a path through space, it can acquire a front (in the direction of its motion) and a back, even though the ball itself has no inherent front or back. Fronts and backs are thus defined relative to body-part projections and typically in relation to the perceiver. It is thus not at all surprising that our bodies, which are the site of our perceptual and motor activities, and which have generic shared structures and part-whole relations, should provide fundamental templates for how we understand and engage our surroundings. Languages around the world all have body-part terms for indicating part-whole and other relational properties of objects, though all languages do not necessarily use exactly the same body-part projections (Brugman 1983; Dodge & Lakoff 2005).

All of this bodily-spatial interaction gives rise to intuitively meaningful patterns of recurring experiences. Such recurring patterns of sensory and motor experiences are what George Lakoff and Mark Johnson named *image schemas*. They comprise structures such as CENTER-PERIPHERY, NEAR-FAR, UP-DOWN, FRONT-BACK, RIGHT-LEFT, CONTAINMENT, SOURCE-PATH-GOAL, OBJECT, and SCALAR INTENSITY (Johnson 1987; Lakoff 1987; Lakoff & Johnson 1999). Image schemas structure a vast range of our mundane bodily experiences of perception and action. They are meaningful affordances (Gibson 1979) for creatures with bodies like ours who inhabit environments like the ones we live in.

Consider, for example, the VERTICALITY schema (or UP-DOWN orientation). Under conditions of normal development, we acquire the ability to stand up in our gravitational field, and we observe objects rising up and falling down to earth. We experience the physical effort (or force) needed to throw something up into the air, or climb up a hill, followed by the fall of the object back to earth or our descent down the hill. We learn the meaning of *up-down* and project that orientation onto physical objects and spatial relations. The top of the hill is *up* and the bottom is *down*. Things *stand up* and *fall down*. What *goes up* must *come down*. The meaning of verticality is not only body oriented, but also oriented as our bodies relate to aspects of our environment with its gravitational field and causal forces. Moreover, it can take on meanings related to our experience of feeling *up* or feeling *down*.

Another significant image schema grounded in our perception and bodily motion is the DEGREE OF INTENSITY schema (or what Johnson 1987 called the SCALARITY or SCALAR INTENSITY schema), which is based on the way our perceptual, proprioceptive, and kinesthetic processes allow us to feel, qualitatively, changes in quality, force, or intensity. Turn a dimmer switch and observe the changing brightness of a light. Feel water go from warm to cool as you add ice. Hear a tone increase or decrease in volume. Experiences like these provide the basis for understanding of scalar changes in the qualities of objects and events.

Besides object, body-part, and spatial relations schemas, there are a number of basic *force* schemas, which arise from our recurring experiences of the way forces act on objects and our bodies, and the way our bodies and other objects exert force on objects. Johnson (1987) described a common set of force schemas, such as COMPULSION, BLOCKAGE, DIVERSION, and ATTRACTION. These schemas share a generic formal structure (a "frame") defined by roles such as *object acted upon, source of force, direction of force, strength of force*, and *resultant state*. Specific force schemas (e.g., TOSS, PRESS, HAMMER, TWIST, POKE, SQUEEZE, MOVE) are then defined by specific values of these general parameter roles.

One highly significant form of forceful interaction is *purposive action*. So, there are a multitude of schemas for basic actions, such as walk, run, hop, skip, jump, saunter, trudge, grasp, pinch, pull, push, throw, eat, and a host of other actions that populate our daily lives. These different actions share a generic action schema or frame structure that has roles for actors, objects acted upon, body part used, direction of action, effort or force of action, manner of action, and so on. There are typically parameters for each of these roles that can be used to specify the particular action being done. So, the THROW schema would have parameter values for the roles such as *actor* (animal, human, or machine), *object* (some moveable object), *body part* (arms, hands, and sometimes trunk and legs), *effort or force* (easy, moderate, or hard), and *direction* (toward some spatial location or an object or person at that location), and so on. Consequently, any category of action can be defined by specifying the relevant parameterizations of the action components or roles. For example, within the entire set of actions of which animals and humans are capable, there is one set that pertains especially to our ability to move our bodies in space. This basic LOCOMOTION schema (Dodge & Lakoff 2005) consists of the following roles or parameters (and their possible parameter values, known as parameterizations):

Mover (e.g., animal, person)

Speed (e.g., slow, medium, fast)

Body part (e.g., feet, legs, trunk)

Gait (e.g., walk, trot, run)

Effort Exerted (e.g., easy, medium, hard)

Walking is a gait of an animal, typically done with the legs (two, or four, or many), feet, and trunk (and sometimes even hands) executed with varying degrees of effort that, in turn, determine the speed. So, the WALK schema, for example, has parameter values something like *actor* (human or animal), *body part used* (legs, feet, and possibly trunk), *object acted upon* (ground), *direction of motion* (some spatial location), and *effort or force* (easy to moderate to difficult). Parameterizations are the particular values given to the different roles or parameters that constitute different aspects of locomotion. For example, the speed parameter can be given the values *slow*, *medium*, or *fast*, such that *run* has the speed-parameter value *fast*, *saunter* has the value *medium*, and *crawl* has the value *slow*.

Dodge and Lakoff (2005) suggest that there are neural mechanisms that generate these particular parameters, and their possible values, for the LOCOMOTION schema, in different types of bodies. In general, for any given action, each role, with its particular parameter values, would need to be realizable by some particular neural architecture and network of neural connections. For example, there are areas of the brain responsible for perceiving spatial location, others that detect motion of objects, and numerous others that control muscle groups necessary for executing certain motor programs, such as those that control the gait of locomotion.

Dodge and Lakoff (2005) point out that different languages can have different ways of representing information about the *manner* and the *path* of locomotion. Two patterns appear to be predominant. One encodes path information in verbs and (optionally) manner information in adverbs. The other encodes manner information in verbs and (optionally) path information in prepositions. English utilizes both of these patterns. For example, English has verbs that indicate manner of motion, such as *stroll, hop, skip, jump, saunter, trudge, tramp, march, jog, gambol,* and *leap,* and the path is indicated by prepositions, as in "John marched *away from* the enemy," "Sue sauntered *toward* the park," and "We jumped *over* the wall and *into* the ditch." English also has verbs capable of coding the direction or path of motion, with the manner represented by adverbs, as in "She *exited* the building *cautiously,*" "I *entered* the room *hurriedly,*" and "The train *departed* the station *slowly.*" According to Dodge and Lakoff, in certain other languages, such as Russian, Chinese, and Ojibwe, manner encoding in the verb predominates, while in Spanish, Japanese, and Turkish, path encoding in the verb predominates.

The important takeaway is that, regardless of which patterns of lexical encoding predominate in a given language, *path* and *manner* are fundamental constituents of actions. They represent universal, cross-cultural structures of bodily acts of locomotion. They are body-based meaning components related to locomotion for creatures with our particular types of bodies, brains, and environments.

Taken by itself, the LOCOMOTION schema does not, as such, specify any particular starting point or goal of the movement. However, we also experience a SOURCE-PATH-GOAL schema based on our observation of motion from a starting point, along a path, and terminating at an endpoint. This schematic SOURCE-PATH-GOAL pattern is present both in the directed motions of physical objects and when we move ourselves from a starting point to a destination. So, our LOCOMOTION schema can be specified by including SOURCE-PATH-GOAL structure to get a form of locomotion that is goal directed toward a destination, as in "He *ran* all the way *from* the barn *to* the house." "Ran" activates the LOCOMOTION schema, while "from . . . to" activates the SOURCE-PATH-GOAL schema, which specifies the starting point, direction, and endpoint of the running activity.

In this way, image schemas provide important meaningful structure to all aspects of our bodily experience. As primary meaning structures, they also have their own *spatial or corporeal logic*. For example, according to the SOURCE-PATH-GOAL schema, if two walkers start at the same place, at the same time, moving along the same path, to the same destination, and if one walks faster than the other, we infer the faster walker will reach the destination first. This inferential relation may seem trivial, but when we later consider our thinking via conceptual metaphor, it goes a long way toward explaining certain inferences we draw about metaphorical objects or events. For example, if we hear that Sophia is *way ahead* of Grayson in getting her bachelor's degree, we reason that she will receive her degree before Grayson, unless something major happens to *block* her progress or to *throw her off track*. It is a simple, but important, logical relation based on image-schematic structure and the PURPOSEFUL ACTIONS ARE JOURNEYS metaphor.

As has often been observed in the cognitive linguistics literature, the CONTAINER schema (consisting minimally of a boundary, interior, exterior, and, optionally, a portal) generates its own spatial logic. An object is either within or outside a container (or it may be passing from one container another). So, if object X is in container A, then it is not outside of container A. Moreover, if object X is in container A, which is then placed within container B, we know (infer) that object X is in container B. This is the transitivity relation in propositional logic, and we learn our intuitive sense of transitivity simply through our repeated experience of spatial containment relations.

In sum, image schemas (1) represent recurring meaningful patterns of our sensory-motor experience, (2) result from the ways our brains and bodies operate within our environment, (3) have their own corporeal or spatial logic, and (4) operate often below our conscious awareness, although they can be reflectively analyzed. Image schemas represent one of our most pervasive and important embodied ways of having, making, and communicating meaning. If any meaningful structures are universal, image schemas would be good candidates. Not that all terms for possible image schemas exist in all languages, but rather that languages the world over seem to operate with many common, shared image schemas. Thus, Dodge and Lakoff (2005) conclude: "We have seen that there is great cross-linguistic diversity in spatial-relations terms and their use of schematic structure. . . . However, we have also seen that these spatial-relations terms can instead be analyzed as complex combinations of more primitive image-schematic structures" (67).

When Lakoff (1987) and Johnson (1987) coined the term "image schema," they sometimes included in their analysis spatial relations, body-part relations, action schemas, and force schemas. Today, for analytical purposes, it may be useful to use "image schemas" for the object and spatial-relations schemas, while treating "action schemas" and "force schemas" as different types. We are here lumping those together as basic patterns of sensory and motor operations rooted in our bodies, but there are clearly characteristics of action and force schemas that may be different from object and spatial image schemas. What is important is to recognize the structural differences of specific schemas, whenever that makes a difference in their analysis and explanation.

6.3   Body-Based Action Concepts

All of our prototypical physical actions—walking, running, lifting, throwing, catching, punching, pulling, grasping, bending, standing, sitting, jumping, and so on—are, of course, done with and by our bodies. We have evolved capacities for carrying off each of these bodily motions, and there are computational neural models of some of these actions (Bailey 1997; Feldman 2006; Lakoff & Narayanan, in press). It should not be surprising, then, that our concepts for these actions are based on some of the same functional neural assemblies that we use to perform those actions physically. In other words, our action concepts are profoundly body based. Our concepts for specific actions are not isolated in some supramodal brain region, but rather are embedded in the same webs of neural connections involved in actually performing those specific actions with our bodies.

There is a history of conceptual theory which insists that our *performing* of a physical action uses different neural ensembles (in different brain regions) than those involved in our *conceptualizing* of that action. Vittorio Gallese and George Lakoff (2005) challenged this traditional disembodied view with examples of how both the action and our conceptualizing of it involve activation of many of the same areas of the brain, especially those involved in perception, bodily movement, and object manipulation. *They describe their view as interactionist, multimodal, and based on cognitive simulation.* To say that a specific action is multimodal is to say that "(1) it is neurally enacted using neural substrates used for both action and perception, and (2) that the modalities of action and perception are integrated at the level of the sensory-motor system itself and not via higher association areas" (459). Multimodal contrasts with supramodal, which requires an association area in the brain that is independent of any sensory or motor areas and is capable of integrating outputs from the sensory and motor systems. Gallese and Lakoff recognize the crucial role of supramodal nodes in neural systems, such as what Antonio Damasio (1999) calls "convergence zones" and what we describe as *heteromodal* cortex in the next chapter. But they argue that our concepts for perception and actions operate for the most part multimodally and not in isolated "supramodal" nodes of neural connections. The multimodal perspective is embodied, for it claims that the same sensory and motor areas that generate actions are engaged as the basis for our conceptualization of those actions. To be clear, both multimodal and supramodal systems and neural architectures play important roles in human cognition, but there are some bodily actions and their correlative action concepts that use exclusively or primarily multimodal systems.

To make these claims about multimodality more concrete, we summarize Gallese and Lakoff's (2005) analysis of the GRASP schema and its correlative concept. Consider, for a moment, how much goes into a simple, almost automatic grasping movement, such as picking up a hammer. First, in preparation for this action, you have to cease any current hand action. Second, you need to locate the hammer visually in peripersonal space. Third, you initiate the grasping action by activating motor synergies such as bending each finger into the appropriate hammer-grasping formation, fingers bent but not too much, and you coordinate the movement of your hand toward the hammer while the fingers begin to close around the handle. Fourth, various body and arm movements must be coordinated into fluid action by the premotor cortex. Fifth, as your hand closes around the handle, you need to exert the right grip pressure to lift the hammer without dropping it. This, and much more! Contrast grasping a sledge hammer with grasping a raw egg. The grip is substantially different and the pressure must be much less, lest you crack the egg.

The general GRASP schema has at least the following parameter components:

1. The role parameters: agent, object, object location, and the action performed.

2. The phase parameters: initial condition, starting phase, central phase, purpose condition, ending phase, final state.

3. The manner parameter (i.e., how the grasping action is performed, including type of grip, strength, and direction of movement).

4. The parameter values (and constraints on them).

5. Body parts used: hands, fingers, and arms.

Gallese and Lakoff (2005) argue that any grasping action can be defined by specifying the values of each of these parameters, and there would be motor programs (rooted in networks of neural connections) related to each parameterization. The various parameters for the GRASP schema would be something like the following:

Agent: an individual

Object: a physical entity with parameters: size, shape, mass, degree of fragility, etc.

Initial Condition: Object location: within peripersonal space.

Starting Phase: Reaching, with direction: toward object location, opening effector.

Central Phase: Closing effector, with force: a function of fragility and mass.

Purpose Condition: Effector encloses object, with manner (a grip determined by parameter values and situational conditions).

Final State: agent in control of object. (Gallese & Lakoff 2005, 467)

This GRASP schema was developed as part of a computational neural model and is meant to be executable in a suitable robotic system. For our purposes, however, the significance of this model is that it (1) reveals the multimodality of the action itself, (2) suggests that our *concept* (here, that of *grasp*) is realized via activations of a selection of neurons that fire in our actual performance of the action, and (3) reveals the simulative character of conceptual thinking. Therefore, our concept *grasp* is multimodal and embodied all the way through. We suggest that a multimodal analysis of this sort would generalize to apply to *all* of our concepts of concrete physical actions.

6.4   The Simulation Theory of Meaning and Thought

The previous abbreviated analysis of the GRASP schema introduces the hypothesis that conceptualization and reasoning are simulative activities. That is, to have a concept of some object or event is to be able to simulate, in the appropriate sensory, motor, and affective areas of our brains and bodies, our interactions (actual and possible) with that object or event. This represents embodied cognition in its deepest sense, insofar as it situates our concepts in our sensory and motor systems, along with the affective responses appropriate to the object or event being conceptualized. This approach has been named the *simulation theory of meaning* (or simulation semantics).

The simulation theory of meaning claims that to understand something (anything) is to be able to run an appropriate embodied simulation of the events and objects being described. Ben Bergen (2012) characterizes the "embodied simulation hypothesis" as asserting that "while we listen to or read sentences, we simulate seeing the scenes and performing the actions that are described. We do so using our motor and perceptual systems, and possibly other brain systems, like those dedicated to emotion" (15). For example, when you read a sentence that starts "She cautiously reached for the doorknob and slowly began to turn it, unsure what might be lurking in the dark hallway . . . ," as you read "cautiously reached for the doorknob," areas of your motor and premotor cortices involved in actually reaching for a doorknob are activated with the appropriate GRASP parameters for that particular object and that particular anxious attitude. In addition, there may be a felt anxiety that builds within you. "The theory proposes that embodied simulation makes use of the same parts of the brain that are dedicated to directly interacting with the world. When we simulate seeing, we use the parts of the brain that allow us to see the world; when we simulate performing actions, the parts of the brain that direct physical action light up" (14–15).

An earlier foundation for simulation semantics was laid by Lawrence Barsalou in his influential essay "Perceptual Symbol Systems" (1999): "Cognition is inherently perceptual, sharing systems with perception at both the cognitive and the neural levels" (577). Consider, for example, what it means to have a concept of a *hammer*. When you hear, read, think of, or imagine a hammer, a flood of sensory-motor-affective activations occur. This includes the visual experiences you have had of the shape of various kinds of hammers (e.g., framing hammers, ball-peen hammers, tack hammers, sledgehammers, wooden mallets, jackhammers, etc.). So, the relevant parts of your visual system are activated. At the same time, different motor programs for using different kinds of hammers (e.g., gripping, raising, aiming, pounding) are activated. There are activations having to do with the weight and heft of various types and sizes of hammers. There may also be associated memories of prior use of hammers. Your multimodal concept *hammer* thus activates functional neural assemblies for these associated perceptions and actions. In a very powerful way, seeing hammers, touching them, thinking about them, imagining them, and hearing or reading about them enacts bodily meanings.

Barsalou (1999) explores the following six characteristics of perceptual concepts, which are concepts of concrete physical objects and actions:

1. *They are records of neural states underlying perception*—"At this level of perceptual analysis, the information represented is relatively qualitative and functional (e.g., the presence or absence of edges, vertices, colors, spatial relations, movements, pain, heat)" (582).

2. *They are schematic and selective*—Attention is selective with respect to what and how it stores an experienced event: "A perceptual symbol is *not* the record of the entire brain state that underlies a perception. Instead, it is only a very small subset that represents a coherent aspect of the state" (583). Perceptual symbols are dynamic, not discrete, and they are compositional.

3. *They are multimodal*—Our perceptual and action concepts involve activations from multiple interrelated perceptual and motor systems in our brains (e.g., visual, tactile, auditory, olfactory, proprioceptive, kinesthetic, motor, etc.).

4. *They are simulation devices*—"Related symbols become organized into a simulator that allows the cognitive system to construct specific simulations of an entity or event in its absence (analogous to the simulations that underlie mental imagery)" (586). To have a concept of a hammer is to have the ability to simulate perceptual and motor interactions (and other factors) with the range of things we denominate as *hammers*. As we will see in chapter 10, the concept of, say, a hammer, is a set of acquired expectancies for meaningful experiential affordances associated with a certain kind of object (here, those things we call hammers).

5. *Perceptual symbols get integrated into larger conceptual frames*—Think how complicated our frame for a *car* is, since it would involve a host of possible perceptual and motor simulations in relation to all of the parts of cars with which we routinely interact. In spite of this complexity, we carry around with us vast stores of knowledge about cars based on our ability to simulate the interactions afforded us by various types of automobiles.

6. *Linguistic symbols develop along with perceptual symbols*—Linguistic symbols come to be associated with a number of relevant perceptual concepts to produce complex semantic fields of related terms: "As simulators for words develop in memory, they become associated with simulators for the entities and events to which they refer" (592).

In sum, embodied simulation semantics hypothesizes that words and other symbols for concrete objects, actions, and events get their meanings through the enactment of sensory, motor, and affective processes associated with certain types of objects and events. When you think about hammers, you are not entertaining a disembodied representation in some inner mental space; rather, you activate a subset of perceptual, motor, and affective experiences associated with hammers and hammering. Cognition and perception are not two ontologically, epistemically, or neurally distinct realms. Instead, cognitive operations (like conceptualization) operate, in part, by activating functional neural assemblies in our sensory and motor systems. Barsalou observes the evolutionary efficiency of such an organization. "Rather than evolving a radically new system of representation, evolution may have developed a linguistic system that extended the power of existing perceptual symbol systems. Through language, humans became able to control simulations in the minds of others, including simulations of mental states" (607).

6.5   Embodied Abstract Concepts: Conceptual Metaphors

It is one thing to claim that our concepts of concrete objects, spatial relations, and bodily actions are embodied in a deep way, but it is quite another to say the same about our abstract concepts, such as mind, thought, knowledge, freedom, right, and good. It makes some intuitive sense that our concrete concepts for objects and events perceived, and actions performed, should use the same neural assemblies that are activated when we actually perceive those kinds of objects and events, and perform those kinds of actions. But what about abstract concepts that appear to be general representations that transcend our particular bodily interactions with and in the world? From a naturalistic perspective, then, the challenge is to explain how our abstract thought and reasoning are shaped by our embodiment, not in the sense that we need a body to think, but in the much deeper sense that our embodiment determines both what and how we think.

Some forty years of cognitive linguistics research has provided considerable evidence that most of our abstract conceptualization and reasoning is done via conceptual metaphors, in which we recruit structures and processes of our basic sensory-motor and social experience to structure our understanding of an abstract target domain. For example, as we saw earlier, one of our most deeply entrenched metaphors for knowing (i.e., KNOWING IS SEEING) appropriates objects, relations, and events from the source domain of vision to conceptualize the target domain of acts of knowing. This thesis was first put forward as a general theory of abstract thought by George Lakoff and Mark Johnson in *Metaphors We Live By* (1980), who coined the term "conceptual metaphor theory." Some of their ideas were anticipated earlier in Michael Reddy's (1979) seminal study of metaphors underlying our conception of communication, and even earlier Max Black (1954–1955) had suggested that metaphoric source domains provided filters for systematically organizing our understanding of a target domain. Conceptual metaphors have been analyzed for languages around the world and in fields and disciplines such as linguistics, philosophy, psychology, science, mathematics, economics, history, law, art, music, theater, and ritual (Kovecses 2010).

What is a conceptual metaphor? Why do we have these metaphors? Where do they come from? *Conceptual metaphors are complex conceptual mappings from a bodily (sensory-motor-affective) or social source domain onto an abstract (nonconcrete) target domain. We acquire these conceptual mappings because of the nature of our brains and bodies (and their capacities for perception, action, and feeling) as they interact with our structured environments. Specific metaphors arise from experienced correlations between entities and events in the source domain and those in the target domain.*

Human evolution has generated brain architectures capable of complex patterns of neural interconnectivity that support concrete and abstract understanding and reasoning. Our brains have evolved to be able to make connections among different functional regions within neural networks (tied to events in our bodies as they engage their surroundings). The precise connections and metaphoric cross-domain mappings are then the result of individual development and learning, which take place in bodies and brains that are structurally similar and engage with similarly structured environments that are at once material, social, and cultural.

In the early years of developing conceptual metaphor theory, Lakoff and Johnson appropriated the term "mapping" (taken from mathematics) to indicate that objects, events, and relations from a sensory-motor source domain are recruited to conceptualize and draw inferences in the (abstract) target domain. Most of the time, the mapping process occurs beneath the level of conscious awareness. For example, the KNOWING IS SEEING metaphor was described as a conceptual mapping of sensory processes involved in vision so as to structure our understanding of acts of knowing (which were commonly taken to be nonphysical, or at least nonconcrete). Their hypothesis was that conceptual metaphors are not based primarily on perceived similarities between parts of the source and target domains. Instead, the mappings were the result of *experienced correlations* between the source and target domains, such as when various acts of seeing objects correlate with gaining knowledge of them.

With recent developments in neuroscience, the term "mapping" now has a neural component because experiential correlations between a source and target domain are realized through patterns of neural connectivity across different functional regions of the brain. Repeated experiential correlations then result in strengthening of the relevant cross-domain patterns of neural connectivity. So, the KNOWING IS SEEING metaphor is seen to be based on neural connections between brain areas responsible for vision and those involved in acts of knowing and action planning (Lakoff 2008; Lakoff & Narayanan, in press).

Joseph Grady's (1997) *primary metaphor hypothesis* attempts to answer the question about how humans acquire the metaphors we use for conceptualization and reasoning, and why we have the particular metaphors we do. Grady explored Lakoff and Johnson's (1980, 1999) claim that conceptual metaphors are "experientially based" on correlations between the source and target domains. Grady developed this into a theory of how we learn certain body- and socially based correlations that constitute a cross-domain mapping (a conceptual metaphor), and then how more complex metaphors can be built up from primary metaphors. A good example of a primary metaphor is the KNOWING IS SEEING metaphor we have been discussing. As sighted children develop, they have recurring visual experiences that afford them understanding or knowledge of certain perceived phenomena. Experientially, there is a felt correlation between seeing (having a visual experience) and gaining understanding and knowledge. This experiential correlation (with its neural underpinning of connections among different functional brain regions) is the basis for acquiring the KNOWING IS SEEING metaphor. Grady suggests that we unreflectively, and mostly unconsciously, acquire a large number of primary metaphors just by living and experiencing correlations across different domains, grounded in the interconnectivity of various functional brain regions. This gives rise to primary metaphors such as MORE IS UP, INTIMACY IS CLOSENESS, AFFECTION IS WARMTH,

TEMPORAL CHANGE IS SPATIAL MOTION, STATES ARE LOCATIONS, HELP IS SUPPORT, PURPOSES ARE DESIRED OBJECTS, DESIRE IS HUNGER, CAUSES ARE PHYSICAL FORCES, RELATIONSHIPS ARE ENCLOSURES, CONTROL IS UP, and UNDERSTANDING IS GRASPING.

More complex metaphors have primary metaphors as their submappings. For example, what Lakoff and Johnson (1999) analyzed as the LOCATION EVENT-STRUCTURE metaphor has for its submappings primary metaphors such as STATES ARE LOCATIONS, CHANGES ARE MOVEMENTS (into or out of bounded regions), CAUSES ARE PHYSICAL FORCES, CAUSATION IS FORCED MOVEMENT, and several more. These submappings are the basis for linguistic utterances such as "The water *went from* cold *to* hot before he knew it," "He *fell into* a depression," "It was the divorce that *pushed him over the edge*," "It took several months for him to *claw his way out* of his listless state," "The slam-dunk *threw the crowd into* a frenzy." Hundreds of cross-cultural analyses of conceptual metaphors over the past four decades reveal how they constitute our understanding of the target domain and guide our inferential reasoning in the target domain (e.g., Dancygier & Sweetser 2014; Gibbs 2008; Kovecses 2010, 2020).

For example, the PURPOSES ARE DESTINATIONS metaphor (as in "I'm *just starting out* to get a Ph.D.," "We're *far from* getting where we need to be on this project," and "She was *so close* to getting promoted") has embedded in the source domain (i.e., bodily motion toward a destination) a SOURCE-PATH-GOAL image-schematic structure that generates the semantics of the source domain and gives rise to inferences (e.g., If I've gone a long way toward my destination, I infer that I'm getting close to it [source-domain inference]. Therefore, via the mapping from source to target, if I've gone a long way toward my desired goal, I infer that I have nearly achieved it [target-domain inference]).

The Neural Theory of Conceptual Metaphor

In *Metaphors We Live By* (1980), Lakoff and Johnson based their analyses primarily on three types of mostly linguistic evidence: (1) *polysemy generalizations*, where the cross-domain mapping could explain why terms with a specific literal sense could be used metaphorically; (2) how *inferences* in the sensory-motor-affective source domain could be recruited for reasoning in the more abstract target domain; and (3) how *novel metaphors* could be explained by their relation to conventionalized conceptual metaphors. Over the subsequent four decades, researchers sought to increase the number of sources of evidence for conceptual metaphor dramatically, and they carried out hundreds of cross-cultural analyses, investigating whether there might be some fairly universal conceptual metaphors and to what extent there may be some cultural differences. As a result of this work, there are now ten or more kinds of evidence for the existence and operation of conceptual metaphor, including polysemy in natural languages, inferential generalizations, novel cases (e.g., poetry, art, advertising), discourse analysis, psychological priming experiments, studies of historical semantic change, multimodal cases, gesture studies, neuroscience, and computational neural modeling (Dancygier & Sweetser 2014; Gibbs 1994; Lakoff & Johnson 1999).

Some of the most exciting and promising research today is the nascent neuroscientific investigation of how our brains make metaphoric cognition possible. While most of this research so far remains tentative and somewhat speculative, it is beginning to provide insight regarding what, from a neural perspective, a conceptual metaphor is and how it structures our understanding, both preconceptual and conceptual. Jerome Feldman (2006), based partly on collaboration with George Lakoff, has suggested some neural modeling that appears to be able to recognize conceptual metaphors and to draw inferences from them. So far, there are only a few relevant neuroimaging studies of metaphor cognition, but George Lakoff and Srini Narayanan (in press) are about to publish a massive and highly ambitious study of the neural foundations of mind, thought, and language, which includes the most extensive research to date on image schemas and conceptual metaphors, to name just two of the many topics they take up. Their account is technical, complex, and dependent on notions they explain earlier in the book, but the basic outline of their neural theory of metaphor can be summarized as follows:

- Mammalian evolution has resulted in the current architecture of the human brain, in which certain brain regions are responsible for specific functions, such as edge detection, motion detection, spatial location in our visual perception, and object manipulation.

- As we will see in the next chapter, these functional areas are often interconnected by more or less dense clusters of neurons. The particular connections, and the strength of those connections, in a person's brain will be the result of their particular experiences over the course of their lifelong development. The neural pathways between the two coactivated regions go both ways between the two regions.

- The more two or more brain regions are coactivated (the more they "fire together"), the stronger the neural connectivity between them, and the more likely that activation of one region will be correlated with activation in the other region. Also, typically the shortest circuit between the two areas gets strengthened, so that a stable pattern of connectivity exists between the two.
- As described earlier, conceptual metaphors are based on experienced correlations between two different domains of experience.
- Neurally, these experiential correlations are realized as time-locked coactivations of functional regions.
- Consequently, when the connections between two functional regions are sufficiently strengthened by repeated coactivation, the basis for the cross-domain mapping that constitutes a metaphor is established.
- However, according to conceptual metaphor theory, the mapping that constitutes the metaphor is directional. That is, the so-called source domain fires first, and the flow of activation is toward the target domain.
- This directionality is the result of what is known as spike-timing dependent plasticity, in which the synaptic connections that fire first in a temporal sequence of activation strengthen the synapses in that direction while weakening the connections coming in the other direction (back from the later-firing neuron). So, as Lakoff and Johnson claimed forty years ago, conceptual metaphors are directional, from source to target. They are not bidirectional, and so they are not based on preexisting similarities that would result if the activations went both ways equally.
- Why should such directionality exist? Lakoff and Narayanan suggest the following answer. Neural clusters activated either for a longer time, more frequently, or from more sources would strengthen the neural activation in the direction from the source-domain neurons to the target-domain neurons. The Purposes Are Destinations metaphor discussed above provides a good example. Out of all the bodily motions we perform, only a subset of these are purposeful. We have many more experiences of random or undirected bodily motion than we do of purposeful action. Therefore, the Motion schema will be activated far more than the Purposeful Action schema. Via spike-timing dependent plasticity, the motion-to-purposeful action direction is strengthened, so that bodily motion becomes the metaphorical source domain, while purposeful action becomes the target.

Lakoff and Narayanan acknowledge that their neural theory of metaphor is speculative at some points, but they point out that something like their view explains the basic phenomena of metaphorical understanding that have been discovered and analyzed over the last four decades of cognitive linguistic research. Our point is that their view builds on recognized neural architectures and processes, so there is a least a plausible neural basis for conceptual metaphor theory—a basis arising from convergent evidence acquired from multiple scientific bodies of research.

To summarize our general account of conceptual metaphor, conceptual metaphor is a principal process of abstract conceptualization and reasoning. It arises from experiential correlations that create cross-domain coactivations (and connections) between the functional neural regions related to the two domains. These conceptual metaphors are body based, getting their meaning and inferential relations from the sensory and motor structures of our bodies and brains as we interact with our environment. Some of them are also rooted in our social relations with others. Metaphor thus recruits sensory, motor, and affective processes for higher cognitive functioning (i.e., conceptualization and reasoning). The source domains of conceptual metaphors come from the nature of our perceptual systems, our motor systems for moving our bodies and manipulating objects, and our intimate interpersonal and social involvements. Each of these source domains are structured by specific image schemas that constitute the basis for conceptual structure, get mapped from source to target, and support the inferences we draw about the target domain based on our knowledge of relations in the source domain. We then understand and reason about the target domain by appropriating the spatial or corporeal logic of the source domain. Finally, it is important to keep in mind that these metaphors are embodied conceptual structures, so they are not merely linguistic and not dependent on language as such. Instead, linguistic metaphor is dependent on prelinguistic metaphor mappings grounded in our sensory, motor, affective, and social experiences. Consequently, conceptual metaphors can be found in all sorts of nonlinguistic meaning making, such as dance, music, painting, sculpture, architecture, spontaneous gesture, and ritual practices (Forceville & Urios-Aparisi 2009; Johnson 2007).

6.6   Conceptual Metaphors for Mind

Conceptual metaphor theory gives us powerful tools for analyzing the fundamentally metaphoric character of our most deeply rooted conceptions of mind, thought, and knowing. In analyzing the conceptual metaphors that Western cultures utilize for conceptualizing thought (and mind), Lakoff and Johnson (1980, 1999) noticed that they were based on bodily activities we engage in every day, such as perceiving, object manipulation, moving our bodies in space, preparing and eating food, and sexual activity. Because these basic bodily activities play such an important role in our knowledge of our world, they have come to be recruited for creating our understanding of various "mental" acts, such as understanding, knowing, and reasoning. For example, once we analyze the cross-domain mapping, we then understand how a term such as *see* can have meanings concerning vision but also for acts of knowing.

The conceptual metaphor here is not the linguistic expression, but rather the system of conceptual mappings by which we appropriate structure and operations from the source domain to understand and reason about the target domain of thinking/knowing. The metaphor is *conceptual*, and then the submappings of the metaphor show up in our linguistic utterances and in other symbolic actions, such as gesture, painting, sculpture, music, dance, architecture, and ritual practice. Here is the partial set of mappings that define the MIND IS A BODY metaphor:

THE MIND IS A BODY Metaphor

Thinking Is Seeing

Ideas Are Objects Seen

Knowing Is Seeing Clearly

Communicating Is Showing

Attempting To Gain Knowing Is Looking For (Searching)

Paying Attention Is Looking At

An Aid To Knowing Is A Light Source

Being Able To Know Is Being Able To See

Being Unable To Know Is Being Unable To See

Impediments To Knowledge Are Impediments To Vision

Knowing From A "Perspective" Is Seeing From A Point Of View

Explaining In Detail Is Drawing A Picture

Directing Attention Is Pointing

Based on this commonplace conventional metaphorical system of mappings, our thought and language manifest the precise details of these submappings, and the polysemous terms they support, as in:

I *see* what you mean. Her argument was *illuminating*. Can you *shed more light* on how quantum physics works? The way I *see* it, we need to get out of Iraq now! He *lost sight* of what he was arguing. I've been *looking for* supporting evidence for embodied cognition theory. Try *looking at it* my way. From my *point of view* there's insufficient evidence to convict. She *pointed out* what was missing in our account of psycho-sexual development. Do I have to *draw you a picture*? I *get the picture*; now I *see*. As I told him what his girlfriend had done, he couldn't even *see what was right in front of his nose*. I'm totally *in the dark* about what happened. Susan was *blind* to everything I was trying to argue.

Another routine yet highly important thing we do with our bodies to gain understanding and knowledge of our surroundings is to move ourselves through space, which allows us to explore our world. This provides the experiential basis for:

The Thinking Is Moving Metaphor

The Mind Is A Body

Thinking Is Bodily Movement

Ideas Are Locations

Reason Is A Force That Moves Us

Careful Thought Is Step-By-Step Movement

Being Unable To Think Is Being Unable To Move

Rational Thought Is Direct, Deliberate, Step-By-Step Motion Forced By Reason

A Line Of Thought Is A Path Of Motion

Thinking About X Is Moving In The Area Around X

Communication Is Guiding

Understanding Is Following

Rethinking Is Going Over A New Path To A Destination

Examples include:

How do we get *from here to there* in this proof? How did you *arrive at* that conclusion? I couldn't *follow* all the *steps* in her argument. Where are you *going* with this? She was *led* to the conclusion that all men are randy dogs. Paul was *forced* to a conclusion he didn't want to acknowledge. I'm not *following* you, so I think you should *walk me through* that again. We *have arrived* at the crucial part of Einstein's argument. My mind has been *racing* all morning. Felix kept *going off in flights* of fancy. Don't get *side-tracked*! *Where are you* in the discussion now? We're just *going over the same ground* again and not getting anywhere. *Slow down*, you're *going too fast* for me. I really can't *keep up with you*. What's the *topic* (from Gr. *topos*, a place) here?

Besides moving our bodies from one location to another, we also use our bodies to manipulate objects for various purposes, thereby giving us an understanding of what objects are like and how they work. This experience gives rise to:

The THINKING IS OBJECT MANIPULATION Metaphor

The Mind Is A Body

Thinking Is Object Manipulation

Ideas Are Manipulable Objects

The Structure Of An Idea Is The Structure Of An Object

Analyzing Ideas Is Taking Apart Objects

Communicating Is Sending An Object

Understanding Is Grasping An Object

Inability To Understand Is Inability To Grasp

Memory Is A Storehouse Of Objects

Remembering Is Retrieval (Or Recall)

Examples include:

> Let's *play with* that idea—*toss it around* for a while. She *turned* the concept of personhood *over* in her mind, considering every *facet* of it. He *broke down* the concept of justice *into its simplest components*. Rachel *constructed* her theory of metaphor *piece by piece*. Jack just couldn't *grasp* what Jill wanted him to do. Oh, I *get it*! *Keep* that concept *firmly* in mind as we proceed with our explanation. Sally *gave* the committee some good ideas to work with. I just don't seem to be *getting* my idea *across to* her. Dr. Know-nothing thinks that good teaching is just *putting* ideas *into* his students' minds—*cramming their heads full* of ideas. What she said *went right over my head*, and when she tried to explain one more time, it all still *went right past me*. Stop, you're *throwing too much at me at once*! When we turned to alternative interpretations of quantum phenomena that really *threw me a curve*.

Another rich and quite interesting metaphorical conception of thinking is built on our experience of finding, preparing, tasting, eating, and digesting food. There is no strong connection between eating and thinking, although tasting things gives some knowledge of their characteristics. However, the things we do to find, process, and consume foods provide a metaphoric model for the kinds of activities we engage in as we think about, evaluate, and communicate our thoughts.

The THINKING IS EATING Metaphor

A Well-Functioning Mind Is A Healthy Body

Ideas Are Food

Acquiring Ideas Is Finding Food

Preparing Ideas To Be Understood Is Food Preparation

Understanding Ideas Is Eating And Digesting

  Interest In Ideas Is Appetite For Food

  Good Ideas Are Healthful Foods

  Helpful Ideas Are Nutritional Foods

  Bad Ideas Are Harmful Foods

  Disturbing Ideas Are Disgusting Foods

  Interesting Ideas Are Appetizing Foods

  Uninteresting Ideas Are Flavorless Foods

  Testing Ideas Is Smelling or Tasting

  Considering Is Chewing

  Accepting Is Swallowing

  Fully Comprehending Is Digesting

  Communicating Is Feeding

Examples include:

> What you've said is certainly *food for thought*. Ludwig had an *insatiable appetite* for philosophy. He couldn't *quench his thirst* for knowledge. Professor Kant *cultivated* the highest ideals in his students. Something *smells fishy* about that theory. Race superiority is a really *rotten* idea. Willard's whole philosophy is just a bunch of *raw* facts, *half-baked* ideas, and *warmed-over* theories. What David Duke said *made me want to puke!* Everything he says is *bullshit*. Let's put that idea *on the back burner* for a while. Let me *chew on* that a bit. We don't have to *spoon-feed* our students, or *sugar-coat* anything. What the senator said *left a bad taste in my mouth. Yuck!* (said of an idea or theory). There's too much here for me to *digest*. Professor Gradgrind just wants his students to *regurgitate* and *spit back to him* everything he's *force-fed* them in class.

These are four of the more important conventional conceptual metaphors for mind and thought that we find in Western cultures. There are others, such as coming up with new ideas as giving birth, as in "Teller was the *father* of the atomic bomb," "Socrates saw the philosopher as the *midwife* who assists others in giving birth to new thoughts," and "That's a really *fecund* hypothesis." These four should suffice to elaborate the major claim of conceptual metaphor theory that our most important abstract concepts are defined by metaphors grounded in aspects of our bodily and social experience. Identifying, analyzing, and giving multiple types of evidence for these four metaphors for thinking does not, of course, prove that our abstract conceptualization is mostly metaphoric. However, over the last four decades, there have been many hundreds of extended analyses of metaphors for abstract concepts in physics, biology, psychology, mathematics, economics, law, politics, religion, fiction, poetry, and all of the arts. These analyses have examined conceptual metaphor underlying languages all over the world (Dancygier & Sweetser 2014; Gibbs 2008; Kovecses 2010, 2020; Lakoff & Johnson 1999).

Reading through these examples engenders that "aha!" awareness of the metaphorical basis of what most of us take, unreflectively, to be our most literal conceptions of mental operations. We simply weren't aware of how important metaphor is in our ability to think abstractly. Nor are we aware of how the bodily source domain generates the polysemy of our concepts (and their linguistic expressions), nor how the source domain generates the patterns of inference we apply in the target domain. It is remarkable how thoroughly, often down to the smallest details, the structure of the source domain generates the exact language we use to talk about our most important abstract concepts. In short, we become more aware of the mostly nonconscious operation of our conventional conceptual metaphors in defining our realities.

It is no accident or mere coincidence that we have these particular metaphors for mind and thought, nor that conceptual metaphors the world over tend to be grounded in aspects of our physical and social experience. We have the conceptual metaphors we do because of the nature of our bodies, brains, and environments. We have the particular metaphors for thinking and knowing that we do because they are based on the kinds of actions we perform with our bodies as we interact meaningfully and knowingly with our surroundings—actions such as perceiving, manipulating objects, moving our bodies, and eating. Those bodily actions are sources of understanding and knowledge, and they have structural features characteristic of our ways of thinking (e.g., we sometimes learn by moving from one place to another; by employing vision to gain highly discriminative understanding of objects, scenes, and events; and by manipulating objects to see what qualities they manifest and how they can be used in various tasks). Therefore, it behooves us to explore the nature and structure of the typical body-based source domains for our conceptual metaphors for knowing. This reveals how profoundly our bodies shape our cognition, from the simplest perceptions up to the most imaginative creations. It also reveals a submerged continent of body-based meaning.

6.7   Conceptual Metaphors for Knowing

We now want to look briefly at some of Western culture's most influential metaphors for knowing in order to see where they come from, how they shape our thinking about knowledge, and in what ways they might be problematic. Interestingly, sometimes our explicit theories of knowing, which insist on literal truth and reject the idea that metaphors are determinative of knowing, are themselves elaborate conceptual metaphors.

Over the entire history of Western philosophy, the most popular and influential metaphor for knowing has been the KNOWING IS SEEING metaphor. According to this common metaphor, acts of knowing are modeled on acts of seeing. IDEAS ARE OBJECTS SEEN, and KNOWING IS SEEING CLEARLY, that is, metaphorically seeing the idea-object in all its detail. The source-to-target mappings give rise to inferences we make. In other words, the logic of the source domain (vision) transfers over to the target domain (knowing), so we reason about the target domain via source-domain inferences. For example, in the source domain, if you see an object in the proper light and without anything obscuring your view, you will see it clearly and objectively. The parallel inference in the target domain is that if you "see" (know) an idea in the proper light of reason, you will know its true meaning and truth conditions.

It makes sense that the KNOWING IS SEEING metaphor provides the most often used framework for theories of knowledge. After all, our highly developed visual capacity, with its fine-detail focus and perception at a distance, is one of the primary means by which we gain knowledge of our world. Not surprisingly, when philosophers have tried to erect theories of knowledge, they have most often appropriated the KNOWING IS SEEING metaphor frame.

For example, Plato built his theory of knowledge upon this metaphor. Recall the Allegory of the Cave, with the contrast between the cave dwellers, chained in the darkness, and the liberated ones who are allowed to exit the darkness of their underground prison into the light of the sun, the source of all knowing. Cave dwellers live, metaphorically, in intellectual darkness, whereas those who have been released from the cave into the light of day possess genuine knowledge of the essences of things. Metaphorically, the light that shines is the light of understanding or reason, and it reveals the nature of what is. Even though the Greeks used the Knowing Is Seeing metaphor, they would never equate perceiving with knowing because they thought that true knowledge must be of the unchanging and eternal forms of things, whereas they viewed the perceptual world as the realm of change and becoming. So, Plato preserves the Greek separation of the perceptual (visible) and intelligible realms while still interpreting the act of knowing metaphorically as a refined quasi-seeing of the eternal forms that make something what it is.

Another famous appropriation of this visual metaphor is René Descartes's positing of an inner theater of mind, in which idea-objects can be illuminated by the "natural light of reason" in such a way that the "mind's eye" can gain certain knowledge of them. Just as seeing an object in the light of day reveals it as it is, likewise, "seeing" an idea-object in the "light" of reason reveals its essential being. Such knowledge results from unerring intuition, by which Descartes means "the conception which an *unclouded* and attentive mind gives us so readily and *distinctly* that we are wholly freed from doubt about that which we understand" (Descartes 1628/1970, 7; emphasis added). Metaphorically, intuition is an instantaneous "seeing" (knowing) of an idea-object, and it "springs from the *light of reason* alone" (7; emphasis added). Descartes insists that knowledge must be based on clear and distinct intuitions, and this requires replacing old, poorly organized, and often confused methods of inquiry with a mathematically modeled method capable of guaranteeing certainty, "for it is very certain that unregulated inquiries and confused *reflections* of this kind only confound *the natural light of reason* and *blind* our mental powers" (9; emphasis added).

Conceptions of knowledge based on the Knowing Is Seeing metaphor have, for the most part, left us with a seriously misleading conception of knowing. They are all predicated on the assumption that the thing known (the Idea As Object, the form, the essence) exists in its fixed and completed form prior to our encounter with it. This is an ontological as well as epistemological claim—the world exists, in its basic structure, in itself, fixed, finished, and mind independent. Hence, acts of knowing have no effect on what is known because ideas or things known are posited as abstract entities in a preexisting and finished realm. The best knowing can do is to mirror those given entities and their relations. Insofar as this takes knowing out of the realm of activity and change, it provides no place for knowing as action transforming our world. This fateful metaphor fails to appreciate that knowing is not merely a representational relation, but rather an active intervention in the course of events in the world, thereby giving them new meaning.

A second metaphor for knowing, based on the THINKING IS OBJECT MANIPU-LATION metaphor, does not fare much better. It metaphorically conceives of the known as a mental "object"—an idea or concept—that can be grasped, turned over for inspection, held in mind, and broken down into its constituent parts (i.e., analyzed). This metaphor emphasizes activity, but it is still a rather blunt instrument for modeling knowing. It lets you "grasp" an idea, share it with another person, manipulate it, and take it apart, but it is woefully thin on the details of how this works. It doesn't explain what "grasping" or "apprehending" an idea involves. How does saying that you've *grasped* an idea explain what that means or how it could give you knowledge? Moreover, say you *analyze* an idea (i.e., break it down into its constitutive parts). How does this constitute knowing?

The inadequacy of the THINKING IS OBJECT MANIPULATION metaphor is not merely an issue about our commonsense folk theories of knowledge being overly simplistic and misguided. The same problem arises in Gottlob Frege's profoundly influential philosophy of mind and language. As one of the progenitors of analytic philosophy, a mathematician and logician, Frege was obsessed with guaranteeing the objectivity of meaning as a basis for objective knowledge. To account for shared meaning and knowledge, Frege (1966) posited an ontological realm (separate from the physical and mental realms) of something like Platonic forms. This unnamed third realm was the locus for concepts, propositions, senses, numbers, functions, and all of the other objective entities that make knowledge possible. His idea was that to understand the meaning of a term is to "grasp" its "sense," which is the public, shared meaning. He thought—quite mistakenly—that our shared concepts had universal senses, which were different from any subjective associations or feelings that might be engendered in a particular person's mind upon entertaining that particular concept. Nobody seemed quite able to explain what "grasping" a sense amounted to, but that didn't stop them from constructing an entire philosophy of language around this idea of objective sense. It is a bit embarrassing to realize that one of the most influential philosophies of mind and language in the twentieth century is apparently grounded on an unexplained metaphor in which "grasp" simply means to understand, without an account of what this grasping action amounts to.

Kant (1781/1968) had divided cognitive judgments into those that are analytic (merely unpacking the meaning of a concept) and those that are synthetic (combining perceptions or concepts into larger meaningful wholes). Consequently, his view provided a basis for analytic philosophy's approach to meaning, thought, and language. Philosophy is called "analytic" because it took the task of the philosopher to be analysis of the meanings of terms and their corresponding concepts (and the relations of those concepts). Analysis is the reduction of a complex "object" (a concept or idea) into its simplest parts. Synthesis is putting simple and complex components together to form complex objects (e.g., concepts and propositions).

The same ontological and epistemological problem encountered with the KNOWING IS SEEING metaphor plagues the KNOWING IS OBJECT MANIPULATION metaphor. The mental objects "grasped," "turned over," "analyzed," or "deconstructed" are, again, preexisting fixed and finished knowledge objects. Our mental manipulations don't alter the ideas; rather, they just allow us to "see" or "grasp" them as they are. There is no way to account for knowing as transformative action. Twentieth-century analytic philosophy eventually became its own best critic when it realized that conceptual meanings cannot be analytically reduced to a finite set of primitive components, such as sense impressions. Instead, concepts are interdefined within semantic fields through complex webs of interconnection (Quine 1960).

The KNOWING IS SEEING and KNOWING IS OBJECT MANIPULATION metaphors are the chief commonplace metaphors we have for knowledge. Neither captures the experience of knowing. Worse, they give us a mistaken ontology—an erroneous view of nature and our place in it, insofar as they posit a set of mind-independent quasi-entities (ideas or concepts) that we "discover" and then analyze. There is no recognition of the transformative nature of human knowing. We suggest that both of these metaphors err in presupposing a completed, stock universe into which acts of knowing enter to capture the essential nature of what is (i.e., what is fixed and final). On the contrary, knowing is an activity for transforming our experience so that we can be more "at home" in the world.

6.8   The Cognitive Unconscious

The conceptual metaphors for mind and knowing represent instances of the way our abstract thinking operates through metaphors that are grounded in our sensory, motor, affective, and social experiences. The evidence for the embodied nature of human abstract conceptualization and reasoning comes from the way the source domains for the metaphors are taken from our physical and social experiences, and then used to constitute the meaning and inference patterns of the more abstract target domain. Even as the mind's process is abstract, creating general understanding that is not limited to specific concrete events, it is fundamentally embodied, grounded in the real experiences of people in their lives. Although the operation of these metaphors can sometimes be self-reflectively attended to, for the most part, they operate automatically and unconsciously to structure our understanding. When we use them to make sense of things, to reason, and to communicate with others, we are typically unaware of them or how they operate. So, conceptual metaphors, along with images, image schemas, action schemas, force schemas, and embodied concrete concepts, are part of the vast nonconscious processes of mind that Lakoff and Johnson (1999) have dubbed *the cognitive unconscious*.

The evidence for the operations of the cognitive unconscious explored in this chapter is primarily linguistic, although there are many types of nonlinguistic evidence available, such as psychological priming experiences (Boroditsky 2011; Gibbs 1994, 2005), spontaneous gesture studies (McNeill 1992, 2005), analysis of multimodal (visual, tactile, etc.) metaphors (Forceville 2009), art (Aksnes 2002; Bhatt 2013), and architecture (Robinson & Pallasmaa 2015). The study of metaphors in language thus illustrates the broader realization that the mind's essential representational patterns, its concepts of the world, are not only found in the propositional logic of linguistic reasoning, linking explicit semantics of language in the ordered syllogisms of rational thought, but also draw on a rich contextual base of meaning that is gained from our many years of physical, embodied experience in the world.

Within the last two decades, the most exciting and extensive evidence has come from neurobiology. We now have neuroimaging technologies and biological evidence from brain and body anatomy that begins to reveal the neural basis of the cognitive unconscious (Feldman 2006; Lakoff & Narayanan, in press). As we investigate from a neurobiology perspective the ways in which the cognitive unconscious gives rise to meaning, values, thoughts, and knowledge, we can start from our animal embeddedness in, and functional engagement with, our surroundings. We have to explore how mammalian learning works and how it has been elaborated in humans, giving rise to remarkable emergent functional capacities for meaning making, motive control of actions, consciousness, and self-aware critical reflection. In chapters 7–10 we explore how the human mammalian need to survive and flourish is regulated by motive controls that shape our perception, reflection, and actions. These motivational structures involve moods and feelings, and they are part of our basic life-maintenance processes. Although these motive controls emerge in relation to initiation of simple actions and later in the formation of personality, they also play a crucial role in all our higher cognitive operations, especially in our ability to make and communicate meaning and to engage in abstract conceptualization and reasoning.

7 The Mind's Anatomy

In the previous chapter we surveyed linguistic evidence for the role of the body in meaning and thought. We looked at both concrete perceptual and motor concepts, but also at more abstract concepts defined by metaphors. However, we need to supplement linguistic analysis with other types of scientific evidence about the workings of the brain and body in the creation and enactment of meaning. In many ways, the appreciation of the embodied process of mind gained from the study of the image-schematic and metaphoric basis of meaning prepares us for the scientific study of the levels of cognitive structure to be found in the brain's network architecture.

Because the vast majority of our cognitive and affective processes operate beneath the level of consciousness, these processes are intrinsically opaque to introspection. Although conscious reflection can provide some key phenomena that must be explained, introspective methods do not allow us to examine the brain events that make these activities possible. We cannot directly experience the firing of neuronal clusters or the secretion of hormones that affect the brain, muscles, and internal organs. We sometimes feel the *effects* of these unconscious processes, but not the *mechanisms* of those processes. Consequently, we need to bring in neuropsychology, and utilize several sources of evidence, including neuroimaging and computational neural modeling, to gain insight into the unconscious, implicit operations of mind.

At the same time, a pragmatic philosophical analysis gives us the context for understanding the scientific evidence. Human cognition is an adaptive function of the whole person. The meaning of any experience is rooted in the needs, desires, and emotional processes that serve the survival and well-being of the organism. Personal experience is grounded in bodily interaction with the world, shaped by our emotional and motivational systems, and then captured in broad concepts (such as image schemas, motor schemas, and implicit metaphors). These are derived from the largely unconscious wealth of physical (sensory, motor, and emotional) experience that has come to be incorporated within our linguistic culture.

In the present chapter we consider biological aspects of cognition that explain this embodied nature of experience and thought. In what follows, a first principle is that *mind can only emerge from a brain that is operating a body that engages its surroundings*. Taking this pragmatic foundation literally, we can read the mind's architecture as we examine the brain's functional networks. These networks provide the road map to the mechanisms for generating the process of mind in the context of the person's world, in both its material and social dimensions.

7.1   Nature's Evolutionary Developmental Process

Over the last several decades, a new evolutionary perspective has emerged in biology to provide insight into the adaptive development of organisms. Genetic control of development is seen not as a simple mechanistic program, but rather a more complex interplay of multiple factors involved in the organism's developmental (epigenetic) self-regulation (Moczek et al. 2015). Developmental plasticity—the self-modification of the organism in the process of environmental experience—is no longer considered a specialized capacity of higher organisms but as a general feature of biological development. The physical changes of the organism that result from its ongoing engagement with its environment reflect the fundamental process by which biological organisms develop and learn. Development continues over the life span of the organism, in each moment, awake or asleep. The continuity from simple to complex forms in evolution becomes clearer as we learn to see it through the lens of evolutionary-developmental theory.

We argue that both the Darwinian and Freudian revolutions in human self-understanding can be integrated into a practical natural philosophy within an evolutionary-developmental perspective. The key assumption is that evolution gives us our basic functional brain regions (e.g., for perception, bodily movements, motivational control, action planning, etc.), and then our individual developmental experiences establish and strengthen specific patterns of neural connectivity among those regions. This evolutionary-developmental process defines who we are and how we experience and understand our world. Patricia Churchland (2002) emphasizes the philosophical importance of the evolutionary component: "We reason and think (and, I would add, feel, perceive, react) with our brains, but our brains are as they are—hence our cognitive faculties are as *they* are—because our brains are the products of biological evolution" (40).

We also need to focus on the individual development component as it progresses over the course of our lives. Each of us is then the result of (1) the biological process of evolution that has resulted in the current brain structure of our uniquely reflective species, and (2) the biological and cultural process of individual development that retraces the evolutionary order to provide the foundation for each person's neuropsychological—and largely unconscious—self-organization. This individual developmental process (ontogenesis) is regulated by genes, but it is not simply deterministic. Biologists describe the developmental process as *epigenesis* because it arises in experience from nongenetic self-organizing influences on gene expression. The epigenetic mechanisms refine and tune the evolved genetic plan to achieve the ontogenetic process that organizes the development of an individual person over her lifetime.

The process of developmental self-regulation has been tuned continuously, but not exactly replaced, by each evolutionary advance in the ontogenetic process (Tucker, Poulsen, & Luu 2015). As a result, we must look to the mind's origins and foundations in the more elementary neural structures of memory and self-control that we have inherited from our ancestors. An overview of the biological foundation for a natural philosophy can be gained by studying the major structures of the vertebrate brain as they have changed in species evolution.

These changes are illustrated by the series of brains used by Niewenhuys, Ten Donkelaar, and Nicholson (1998) to illustrate the progression from relatively primitive toward more complex, and generally more recently evolved, species (see figure 7.1). The major divisions of the brain have been conserved throughout the vertebrate lineage. Roughly, it may help to think of the evolution of brain anatomy across species as a process in which new functional brain architectures are gradually added to previous brain regions, in such a way that the new architectures take over some (but not all) functions previously carried out in earlier established regions, at the same time as they introduce new functional capacities.

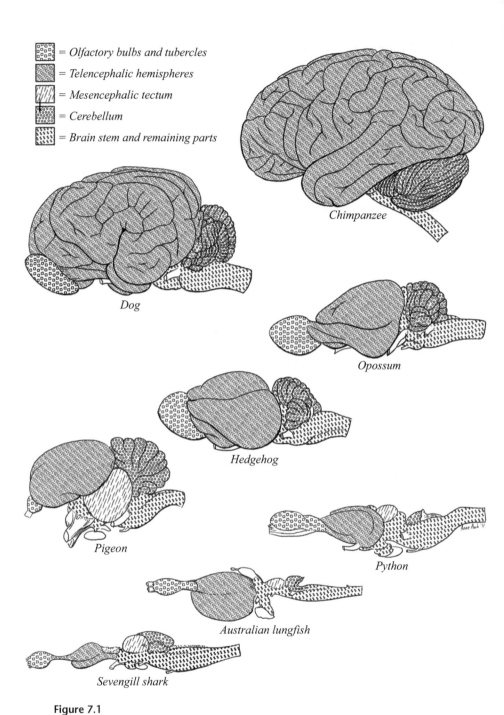

= Olfactory bulbs and tubercles
= Telencephalic hemispheres
= Mesencephalic tectum
= Cerebellum
= Brain stem and remaining parts

Chimpanzee

Dog

Opossum

Hedgehog

Pigeon

Python

Australian lungfish

Sevengill shark

Figure 7.1
The elaboration of the telencephalic hemispheres in evolution can be illustrated roughly through comparison of extant species of varying levels of complexity (after Niewenhuys, Ten Donkelaar, & Nicholson 1998).

Now, can this biology be philosophically meaningful? We may be able to imagine the experience of a chimpanzee or a dog, but birds and snakes not so much. Yet, for the neuroscientist the general architecture of the vertebrate brain has remained through each of the many mutations that supported the many speciation events of our natural heritage. For example, although the telencephalic (endbrain) hemispheres are large and certainly important in humans and other big primates, they only function through the support of more elementary subcortical levels of neural organization, including the diencephalon (interbrain), mesencephalon (midbrain), and rhombencephalon (brain stem) (see figures 7.1 and 7.2).

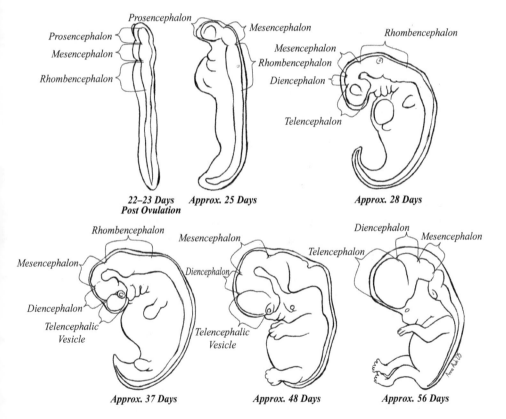

Figure 7.2
Human embryogenesis proceeds through the progressive elaboration of the vertebrate plan as programmed in the embryo's genome and as actualized in the epigenetic, self-organizing, developmental process.

By studying the continuity of these evolved forms, we see that as evolution has given rise to more complex "higher" functional neural organizations, these emergent functions typically rely upon the continued operation of lower-level functional structures, rather than leaving them behind. In fact, the most basic controls on neural activity in the human brain are those of the brain-stem activating system, which are then elaborated in unique ways at subcortical, limbic, and neocortical levels.

There is, of course, a prodigious history of vertebrate neural evolution that is difficult to grasp with the simple overview presented here. Yet, understanding the embodied mind within a modern natural philosophy must begin with some appreciation of evolution. From that perspective, we discover the continuity between the elemental themes of neural control systems in simpler vertebrates and those operating in more complex "higher" vertebrates.

Organismic self-regulation—the essential requirement for biological development—requires the regulation of neural activity over time. This becomes important first in forming and eliminating connections in embryonic circuits in early neural development. It continues in regulating elementary memory capacities in the neural ensembles of both subcortical circuits and the primitive vertebrate cortex. As we will see in chapters 8–10, this control of neural activity in time first regulates the assembly of the neural architecture in embryogenesis, and these same activity control mechanisms then regulate fundamental features of attention and memory in the adaptive self-regulation of experience and behavior throughout life. In this sense, *knowing is a direct continuation of the process of growing*. In this literal sense, the mind is embodied from its first potentiality in the neural structures of the embryo, all the way up to the most complex cognitive capacities of adult functioning.

Therefore, to build an intuitive understanding of the mind's developmental anatomy, it is useful to examine these processes in the early stages of life as each person differentiates from a single cell into a sentient infant with the nascent capacity of mind. The genetic plan for an individual brain is a developmental plan, meaning it must unfold in an ordered sequence. As a result of the myriad ontogenetic mutations of vertebrate evolution, the human brain is organized on the phyletic plan (compare figure 7.1 to figure 7.2), with the large and differentiated telencephalon (end brain) emerging as the developmental outcome of the more fundamental diencephalic (interbrain), mesencephalic (midbrain), and rhombencephalic (brain stem) architectures.

Psychologists and philosophers typically assume such anatomical details are the purview of biologists and not particularly relevant to their studies of the mind and knowing. But modern scientific research reveals that the process of neural differentiation continues throughout life and is synonymous with the organization of concepts and memories. Memories are formed as certain synapses (connections between neurons) are stabilized and others are eliminated. The control of neural activity in time, through elementary mechanisms such as habituation and sensitization, provides an essential substrate for more abstract levels of conceptual development. We understand now that neural growth is identical with the process of experience. *Cognition is neural development continued in the present moment* (Tucker & Luu 2012).

It may be unintuitive to consider each thought and feeling as a process of brain growth. Yet, this seems to be the reality of mind as a biological process, embodied in the brain's synaptic architecture, carried out in each instance through mechanisms of neural differentiation and growth. The mind increases its information complexity in the same way that life spontaneously maintains and increases its complexity of tissue organization.

The next two sections outline the basic architecture of the human brain that we will use to understand how the mind is embodied, literally, in its neural networks. These dense summaries must be studied in relation to the figures, rather than simply read, to be grasped fully. The reward for this study is a clear grasp of the brain's organization that has only become apparent in the last several decades.

7.2   Levels of Neural Organization

The adult human brain (figure 7.3) is not radically different in appearance from that of the chimpanzee brain (as in figure 7.1). As shown in figure 7.4, the large cerebral hemispheres wrap around the subcortical structures, and these retain the basic organization of vertebrates generally. The base is formed by the brain stem and midbrain, the interbrain is formed by the thalamus and hypothalamus (diencephalon), and the highest level is formed by the telencephalic hemispheres (cerebral cortex and basal ganglia).

Much of the difficulty in understanding neuroanatomy comes from the highly folded nature of the brain, causing the functional regions to be

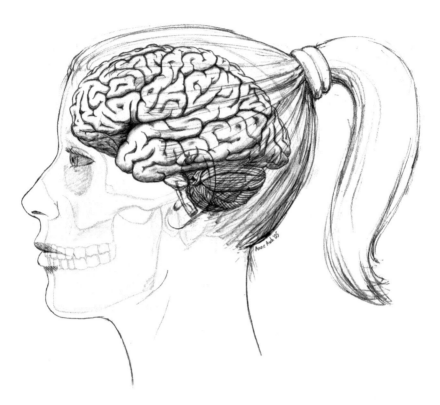

Figure 7.3
The adult human brain, seen from the left side.

positioned in unintuitive ways. This folding seems to have occurred as the cerebral cortex expanded during evolution, folded over the subcortex and brain stem, and gradually became more wrinkled in order to fit its large two-dimensional sheet into a skull that can pass through the birth canal. The progression of forms in figure 7.1 gives a good sense of the position of the telencephalon at simpler stages of evolution, and the embryological progression of forms in figure 7.2 shows how it is built on the ordered circuitry in each child's development.

Without immersing ourselves in the intimidating details of the complex relations within and among various levels of neural organization, we at least need to understand the functions performed by these various neural architectures. Because animals need ways to monitor their internal body states, and they also need ways to sample and interact with their

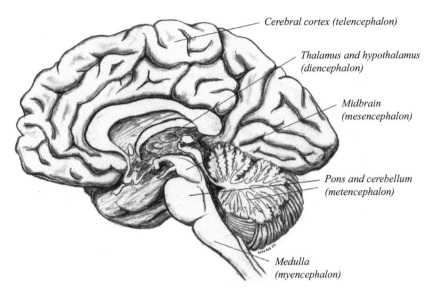

Cerebral cortex (telencephalon)

Thalamus and hypothalamus
(diencephalon)

Midbrain
(mesencephalon)

Pons and cerebellum
(metencephalon)

Medulla
(myencephalon)

Figure 7.4
The human brain, seen from the medial wall of the right hemisphere.

surrounding environments, it is to be expected that they have evolved neural architectures to carry out these vital functions. Consequently, we find neural structures at different levels of organization for *visceral* (i.e., body-state monitoring) and for *somatic* (i.e., perceiving and acting in the environment) functioning. This division between visceral control at the limbic core of the cerebral hemisphere and somatic control in the sensory and motor neocortex is illustrated in figure 7.5.

Figure 7.5 is a schematic illustration of the right hemisphere (with the left hemisphere invisible), just like figure 7.4 only in cartoon form. The outer oval encompasses the whole hemisphere, which is functionally organized with output (motor control of action) in the frontal areas (on the left) and input (perceptual organization of sensation) in the posterior areas (on the right). The inner oval describes the limbic cortex, at the inner core of the hemisphere (*limbus* means border, and this is the inner boarder of the cortex with the subcortical areas). This limbic cortex also has output and input functions, specifically for the body's internal visceral control.

This limbic specialization—for visceromotor functions in the dorsal division versus viscerosensory functions in the ventral division—may be a clue to the organization of the limbic system's control of the dorsal and ventral

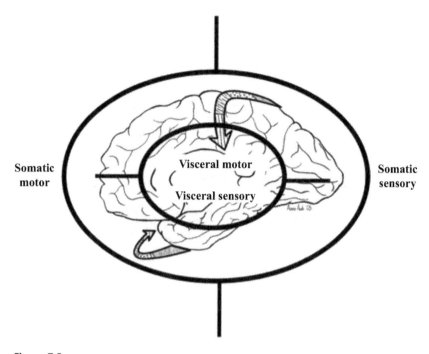

Figure 7.5

The limbic core of the brain regulates the internal visceral state, with the output or visceromotor function in the dorsal limbic division, and the input or viscerosensory function in the ventral limbic division. The more differentiated neocortex provides the interface with somatic functions of body sensation and action, with motor control in anterior regions and sensory control in more posterior regions.

divisions of the neocortex generally. This dorsal and ventral division of the hemisphere, organized by the visceromotor limbic functions dorsally and viscerosensory limbic functions ventrally, will become a key organizing principle as we study the operations of mind in chapters 8–11.

This dorsal/ventral specialization for the limbic visceral function suggests that internal feelings have a sensory aspect (viscerosensory) and also a visceral action aspect (visceromotor). But already the brain architecture has become complex enough that it is not easy to understand why the input and output functions of the brain are separated in this way, with *somatic* (musculoskeletal and external sensory) contact with the world separated between frontal and posterior divisions, whereas *visceral* internal control is separated between dorsal (top) and ventral (bottom) divisions.

As always in biology (and now philosophy), evolution explains things. We can see the basic sensory, motor, and visceral organization in simpler vertebrates (figure 7.1) and in the early stages of our own brain growth (figure 7.2). As each major transformation of vertebrate neural architecture appeared (mesencephalic, diencephalic, telencephalic), there was increasing complexity in both sensory processing and motor processing, and in the control of these in relation to visceral homeostasis. The mammalian brain underwent a very complex transformation, in which sensory and motor capacities were no longer segregated as clearly as they were in simpler vertebrates (Tucker & Luu 2012). Although the specifics of this mammalian mutation are indeed complex and still a matter of debate among neuroscientists, we can see the general results in the organization of figure 7.5, and we can readily visualize the evolutionary transformation through examining the full human neuraxis in figure 7.6.

The *neuraxis* is the whole nervous system, from head to tail. As shown in figure 7.6, the lower levels of the human neuraxis, as in the spinal cord, retain the simpler functional organization of early vertebrates. The sensory functions are organized in the *dorsal* division of the spinal cord (and brain stem), and the motor functions are organized in the *ventral* division. This separation was transformed with the remarkable mutations that created the mammalian neocortex, such that in the telencephalic hemispheres the dorsal division took on motor as well as sensory functions (Tucker & Luu 2012). As this happened, the visceromotor functions (expressing internal needs and urges) seemed to have become integral to the dorsal division of the mammalian hemisphere, as suggested in figure 7.5. In figure 7.6 the *cingulate gyrus* is the major cortical landmark of the dorsal limbic division, and its visceromotor base is now essential to both somatic motor (frontal cortex) and somatic sensory (posterior cortex) functions. The ventral limbic viscerosensory function is illustrated in figure 7.6 by the *insular region* of the limbic cortex, and the viscerosensory functions also span both somatic motor functions (in the frontal lobes) and somatic sensory functions (in the posterior brain).

As we study the implications of the brain's architecture for human cognitive function in chapters 8–11, we will see that the embodied mind reflects both domains of its neural substrate, including not only the somatic functions that allow us to sample the world through our senses and act upon it through our muscles, but also the visceral functions that provide unique

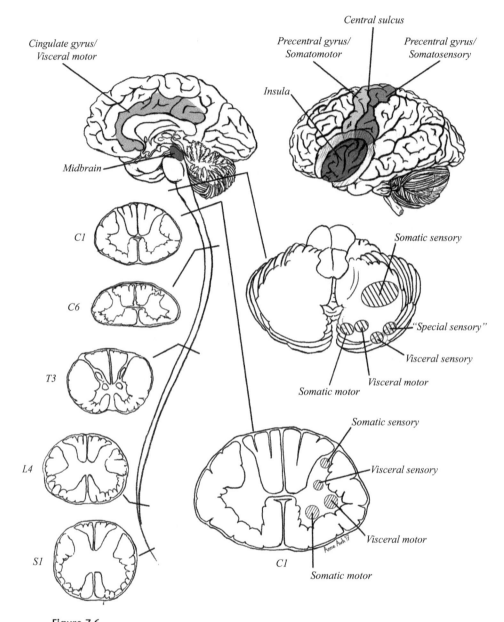

Figure 7.6
In the vertebrate spinal cord and in the brain, the general plan has sensory systems located dorsally (toward the back) and motor systems located ventrally (toward the front).

forms of motive control to the dorsal and ventral divisions of the limbic cortex and neocortex.

To appreciate these biological foundations of mind, the naturalistic philosopher must become a neuropsychologist in order to gain basic intuitions for navigating the mind's landscape. Even with this rough overview, already we see key elements in the functional organization of the brain that will have clear implications for the organization of mind. As we look at the connectivity of the regions of the cortex, the major pathways of the brain connect between the primary somatic cortices (which interface sensation and action with the outside world) and the visceral limbic cortices (which interface the brain's memory and cognition with internal bodily states and needs). What we will discover in the pages ahead is that the mind's cognitive representations (concepts) appear to be organized similarly, contacting the world through the operations of sensation and action, and gaining personal meaning through evaluation at the limbic core.

7.3   The Integration of Mind at the Limbic Core

Within this general organization, there are specific patterns of neural connections in the limbic cortex, and these form the base for the patterns of connections in the sensory, motor, and *association* regions of neocortex of the cerebral hemispheres. These are called association regions because they are not dedicated to sensory, motor, or visceral functions specifically, but rather have connections to multiple modalities of sensation and action. Because these functional regions integrate multiple different perceptual and action processes, they are also known as *heteromodal* cortex. There is a regular organization of the neural connections that define the cognitive function of each of the divisions of the hemisphere, and these have been traced carefully in primates over the last half century. More recently, human cortical connectivity has been examined with advanced methods of neuroimaging, and these findings generally conform to the nonhuman primate (monkey) studies. The result is that we understand the general wiring diagram of the human brain. As with an engineering drawing, once we know the wiring diagram, only certain functional interpretations remain viable.

A key component of the naturalistic account of cognition that we are developing is the discovery of four major types of cortex, here illustrated in the top of figure 7.7, which shows the medial wall of the right hemisphere.

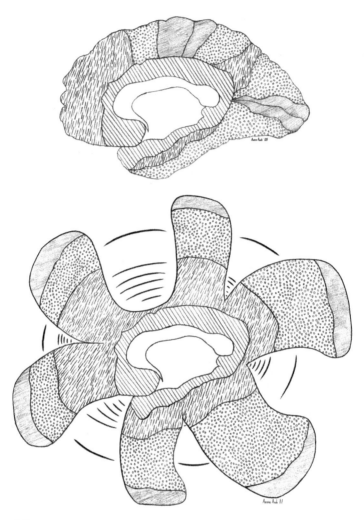

Figure 7.7

Cartoon of the cortical and limbic architecture of the cerebral hemisphere. *Top*: Right hemisphere seen from the medial wall, similar to figures 7.3 and 7.4. This includes primary sensory and motor cortices (*shaded*), secondary (unimodal) sensory and motor cortices (*stippled*), heteromodal association cortex (*dashed*), and limbic cortex (*striped*). *Bottom*: Cartoon of the cortical pathways unfolded, showing that each sensory pathway (illustrated by the four pathways at the *right*) and each motor pathway (illustrated by the two pathways at the *left*) includes primary, secondary (unimodal), heteromodal, and limbic networks, with both dorsal and ventral streams for each. The interconnections among pathways are nonexistent between primary sensory and motor cortices (with the exception of somatosensory and motor), are sparse for secondary cortices, and are densest for heteromodal and limbic cortex.

We will be emphasizing the crucial role of these different cortical structures in perception, movement, and other higher cognitive functions. The four distinct cortical architectures are: (1) the visceral limbic cortex at the hemispheric core (cortical representation of homeostasis, here marked with stripes); (2) the heteromodal association areas (multiple sensory modality and motor connections, dashed); (3) unimodal (or secondary) association area (more complex, association cortex, but linked directly to a sensory modality or motor control, stippled); and (4) primary sensory and motor cortices (the first cortical representation of the sensory data, shaded). These are the component platforms of the mind's anatomy. We cannot easily introspect their contents, but now that we have discovered them, we can reasonably infer their contributions to the unconscious neural base of experience.

It is necessary to understand what each of these four types of structure does because they become the basis of our concepts and our capacity for abstraction (explained in chapters 9 and 10). These four types of cortical structure make it possible for us to sense our surroundings, combine different types of sensory input, and project actions into the world. In sense perception, for example, elements within a single sensory mode are registered in the *primary sensory cortex*, combined into a unified image within *unimodal (or secondary) association areas*, combined with input from other sensory modes in *heteromodal association areas*, and set within a holistic, affect-rich context in the *visceral limbic cortex*. In the bottom half of figure 7.7 is an illustration in which the organization of these cortical regions is presented in schematic form. The interconnections of the cortex are clearly organized by these divisions.

Each *sensory pathway* for input proceeds from primary sensory cortex to unimodal (secondary) association to heteromodal association to limbic cortex. The limbic cortex, although not as differentiated as the lateral neocortex of the hemisphere, remains important in regulating the function of the entire hemisphere (García-Cabezas, Zikopoulos, & Barbas 2019). For example, consider a simple case of visual perception. In the primary sensory cortex we have functional neural clusters that register points, lines, edges, orientations, center-surround configurations, and so on. These visual elements have to be unified in unimodal cortex as an image (not necessarily conscious). However, visual perception is not just visual because it also includes inputs from other sensory modalities, such as tactile, olfactory, and auditory inputs. These inputs combine in heteromodal association areas, giving us an experience with multisensory qualities. Finally, the perceived object is located in space relative to a broader adaptive context—such as whether the thing is good to eat—through activations in limbic cortex. In perception, therefore, the within-pathway connections dominate the wiring of the cortex, such that the primary traffic comes from the outside world through the primary sensory cortex and toward the unimodal (secondary), heteromodal, and limbic cortices.

In contrast, each *motor pathway* of the frontal lobe proceeds in the reverse direction, from limbic to heteromodal motor to premotor motor to primary motor cortex. So, in the case of action, we start with a general action plan projected toward the world (this begins with a need, urge, or impulse), and then modulate the precise movements needed to engage the particular details of the environment in which we are acting.

In addition to these overall dimensions of brain organization (limbic core versus neocortical interface with world; anterior motor versus posterior sensory), it is important to recognize how the limbic division between dorsal and ventral subsystems figures in psychological function. These are the cognitive elaborations of the visceromotor and viscerosensory functions. In perception, the dorsal system (top of the upright human brain) has been shown to process *configurations* of information, such as in orienting the body in space or planning actions in a spatial context (Goodale & Milner 1992; Milner & Goodale 2008). It is often called the *where* pathway because it processes spatial information. The dorsal division of the brain has all the network levels dedicated to processing the spatial context for experience.

The ventral system (bottom of the upright human brain) is specialized for perceiving specific and enduring *objects* in the world, regardless of the context in which they appear. This has therefore been called the *what* pathway. The ventral division also includes all the network levels, illustrated in schematic form in figure 7.7.

The dorsal and ventral divisions for action are complex, but they parallel these perceptual specializations, with unique forms of action regulation in dorsal and ventral divisions of the hemisphere, as we will see in chapters 8–11.

The main connectivity of the cortex is organized within the pathways illustrated by the six flaps in figure 7.7. There are also connections between pathways. These are largely restricted to communication between similar levels of cortical differentiation, for example secondary association to secondary association, heteromodal to heteromodal, and limbic to limbic. This pattern of connectivity provides additional clues to the segmented strata of the mind's architecture that lurks behind each conscious experience.

Because the primary sensory and motor cortices connect to the rest of the cortex only through their secondary association cortices, the implication is that the sensory and motor interface with the world is achieved through cortical "islands" that mediate between the world (as linked through the sensory or motor traffic of thalamus or spinal cord) and the more integrative functions of association and limbic cortices. The connectivity between cortical pathways increases in density toward the limbic core. There are thus minimal interpathway connections for secondary cortices, moderate connections for the heteromodal association cortex, and very dense connections among limbic regions.

One important implication of this brain structure is that levels with the greatest density of interconnections (such as the limbic cortex) provide a more holistic sense of the context. Heteromodal association areas also provide connections across multiple perceptual modes. But then primary and secondary sensory areas tend to be more modular, since they have only sparse connections with other brain regions. Consequently, the greatest modularity is found in the primary sensory areas and to a lesser degree in the unimodal (secondary) sensory association areas, while we can infer from its connectional anatomy that the limbic cortex is best organized to generate a holistic, affect-rich sense of context.

Think, for example, of seeing a baseball. That ball is situated within a spatial context (via the dorsal limbic cortex). You seeing it as a unified image is the result of your ability to recognize lines, edges, colors, and so on (via primary [unimodal] sensory cortices). Other aspects of the ball—the feel of it in your fingers, its distinctive rawhide odor, and its heft in your hand—arise via secondary heteromodal association cortices. Your personal experiences with a baseball, whether rich or fleeting, are recruited to bring the feeling tone to the experience. The result is the complex, multimodal experience of seeing and feeling a baseball in your hand. Moreover, even when you are only seeing the ball without handling it, neural clusters involved in touching and manipulating the ball are activated as part of your visual experience.

7.4   Interpreting the Mind's Connectional Architecture:
How Expectations Shape Perception

One of the central hypotheses of this book is that the brain architecture we describe as different levels of neural structure and organization constitutes the basic architecture of mind. This may not be obvious now, but in chapters 9–11 we explain how these four types of cortical structures make abstraction possible and create our concepts. Here, in this present chapter, we get a glimpse of how neural connectivity implies conceptual organization. First, we see there are four levels in the neural organization of mind, each of which is mostly coherent with itself (through the level-specific interpathway connections). There are indeed specific modules for both the primary sensory cortices and motor cortex, with unique input-output relations to sensation and action. These interface with the rest of the brain through the secondary (unidmodal) association cortices, which provide more conceptual rather than sensory or motor transduction functions. In vision, for example, where the primary cortex performs low-level organization of lines and receptive fields, the secondary association cortices begin to provide more meaningful representations of motion, spatial patterns, and the objects of perception.

These more integrated representations are then restricted to heteromodal association cortices, particularly the limbic cortices. The heteromodal association areas are important integration or *convergence* zones (Damasio, Tranel, & Damasio 1990) in both the posterior brain (parietal and temporal perceptual cortices) and the anterior brain (the motor evaluation and planning cortices). The limbic areas are the most densely interconnected across regions, implying that—with their visceral evaluative and motive functions—these networks must form the integrative core of the mind, not only interfacing concepts with bodily needs, but also organizing even the more abstract conceptual operations as a function of personal values and motives.

This last interpretation is a somewhat paradoxical implication of the brain's connectional architecture. We tend to think of the visceral level as concrete and simple, not as integrative and abstract. If something makes you disgusted, for example, this is not an abstract experience: you just want to throw up. Yet, the dense interconnections among the limbic networks at the core of the brain, compared to the greater isolation and modularity of the sensory and motor cortices, imply that our visceral feelings actually form the integrative base of experience. We form even complex intellectual evaluations with a base at the gut level. Because this integrative level is where our basic values operate and where emotions come into play, we can begin to understand the implications of this cortical connectional anatomy for the emotional component of abstract concepts.

Another principle comes from studying the connections that link the visceral limbic base to heteromodal association, secondary association, and primary sensory and motor cortices. The bidirectional information flow (limbic toward external world and external world toward limbic) is an important feature of cortical architecture. This bidirectional pattern is surprising if we assume only that perceptual data comes *into* the cortex and motor data goes *out* again. Instead, neural processing is bidirectional for both perception and action. We reach out with our expectancies in order to see. We imagine (and feel) the effects of our movements in order to act.

It is not that an urge or motive arises de novo in limbic regions, proceeds to planning in the frontal association cortex, and then is articulated in movement in the motor cortex. Rather, the bidirectional flow of information means that the constraints of the action plan from motor cortex always feed back to the planning and motivation centers to shape the ongoing process of developing action.

Recall our prior example of the schema for grasping as it shapes our use of a hammer. Based on many prior experiences with hammers of various kinds, we developed anticipatory action plans for using hammers that guide and modulate each phase of a particular action. As you launch the grasping action, your prefigured grasp schema needs to be continuously modified to meet the specific demands of *this* particular hammer. There is constant visual and tactile feedback as you form the right grip. If the hammer turns out to weigh considerably more than you expected, you must calibrate the force of the grip and activate the right muscle groups to grasp and raise the hammer. You don't just *see* the hammer and then begin to construct the appropriate action plan. Rather, you initiate the action and then tune the precise details of the action, using feedback gathered at the sensory and motor interface with your present situation. Actions must be continually guided by the feelings of our bodies in movement.

Most neuroscientists have typically assumed that sensory data comes in through primary sensory cortices and is then interpreted in more complex ways in association areas in the posterior brain. They described this direction of traffic as "feedforward," as if it is the primary flow of information. Instead, judging by the anatomy, the high density of neural traffic in the reverse direction implies that memories and expectancies in association areas are shaping the process of perception as much as the sensory input does. In cybernetic (control theory) terms, this shaping of perception by expectancy would actually be the feedforward form of control, setting out what's expected to be perceived before the sensory data can be meaningfully processed. We have to expect before we can see.

This discovery is of the utmost significance for our understanding of experience, action, and knowing, and we will study it carefully in chapters 8–11. The emerging view of the functional traffic across the linked networks of the cortex thus reverses the assumptions of most philosophers and neuroscientists. Contrary to what empiricist philosophers (Hobbes, Locke, Hume, et al.) assumed, the incoming messages from the senses are not the primary input that is then analyzed and evaluated in thought. Contrary to what many neuroscientists assume, the sensory data is not feedforward information. Instead, the feedforward control is actually the *expectancy* formed at the core of the brain, in limbic and association cortices.

This expectancy is then the beginning of the perceptual process. It projects a hypothesis of sorts, in the form of a tentative percept, to pattern the sensory input. The sensory data then serves up error signals reflecting how the expectancy matches the data of the world, or not. Human perception is an ongoing process in which we acquire expectancies based on our prior experiences, and these then guide our exploratory actions in our world. Importantly, expectancies are grounded in visceral evaluations in limbic networks. We then test these expectancies against each present encounter with the surroundings, adjusting them in order to accommodate discrepancies. *There is no such thing as pure, objective perception, since all perception results from prior expectancies and the values latent within them.*

Although such an interpretation may seem surprising to our naive view of the mind's perceptual process, Dewey anticipated it in his important article "The Reflex Arc Concept in Psychology" (1896). More recently, the key role of expectancy in organizing experience has been proposed by at least one psychologist who has studied human perception carefully. Roger Shepard (1984) concluded that perception is essentially hallucination constrained by the sensory data. In perception we project an expectancy about what we will encounter in some future experience (this is the "hallucination"), which we then test against the light striking the retina (this is the constraint by the sensory data). Human perception is organized by experience, by the structures of the mind that have become educated to construct meaning from the sense data. In psychological terms, the bidirectional architecture of cortical pathways—running from visceral to somatic networks—emphasizes the importance of prior knowledge in organizing the process of perception. Consequently, prior experiences with some object, person, or event will build up expectations that shape our current and future perceptions and actions.

For example, say you run into your physician at the movie theater. Initially, you might not recognize her because previous interactions with her have always been within the formal setting of a doctor's office. You did not expect to see her in a movie theater. Your perception is quick in the context of her office because your visual (and cognitive) system is primed by that context. It is slow and effortful in the movie theatre because you have little expectancy to prepare the mind for the perception.

Both philosophical and psychological approaches to the mind have long emphasized perception as the bedrock of objectivity, upon which more complex interpretations must be built. This is the *dogma of immaculate perception*. Even a cursory examination of the neural networks of perception leads to the implication that perception cannot simply be read from sense data. Each person's neural network constitutes a life history of perceptual experience that provides well-organized expectancies for the patterns to be assembled from the senses. Experience is not only embodied but *embedded* within an environment. In both ways it is always expectant, reaching out to predict the events in the world. This is an insight that figures importantly in the recent theoretical advances in human neuroscience to be examined in the next chapter. As we will see, the implications of this process of expectancy followed by testing are profound for understanding the process of knowing.

This chapter has introduced some fairly complex outlines of the brain's organization. It is not a simple matter of understanding uniquely human brain organization. Rather, the biologist (and the student of natural philosophy) must appreciate the course of vertebrate evolution to make sense of the architecture that has resulted in our human neural mechanisms. The reward from a careful study of this hierarchic, evolved neuroanatomy is that we can see what must be the mind's organization. Our conclusions, for example, about the limbic base of experience, and the need for motivated expectancies to shape perception, follow directly from reading the processing implications of the way the brain's connections are organized. Neural architecture reveals the functional organization of mind. And it helps to explain who we are.

Reading the mind's functional capacities from its connected architecture is a daunting challenge of today's neuroscience. In the next chapter we will see that this challenge has been made somewhat more tractable by advances in computational modeling of intelligence. The remarkable scientific convergence in today's cognitive neuroscience has occurred because we have adopted the lessons of the brain—how massive connections are made among simple neurons—in building artificial intelligence (AI). In turn, because we can form intelligent perceptions and decisions through the mathematics of artificial neural networks in AI, we can now draw on more explicit intuitions as we read the connections of the brain's anatomy. In the next chapter we review the growth of the core concepts of AI over the last seventy or eighty years, and then explain how these concepts reveal how the mind's structure emerges from the brain's architecture.

8   How Information Is Captured by Neural Networks

We have proposed that the structure of the brain's networks implies the structure of mind. In order to appreciate this literal interpretation, it is important to address the basic question of how the brain's interconnections allow us to represent information. The advances in mapping the connections of the brain reveal how functions (e.g., edge detection in vision, visual motion tracking, registering pitch changes, hearing a certain timbre, initiating hand actions, etc.) arise from patterns of extensive connectivity among neurons. This is not a straightforward lesson. The human brain has a massive cerebral cortex, with tens of billions of simple cells, neurons, and astrocytes, with trillions of synaptic interconnections. We are used to thinking of communication between sophisticated information processors, such as people or computers. Instead, the connections of the brain are between relatively simple cells, often described as "dumb neurons." Advances in artificial intelligence (AI) have recently shown how cognitive capacities, such as recognizing individual faces or finding cats in internet videos, can be achieved by artificial neural networks (ANNs) made up of similar massive networks of dumb neurons.

We introduced the concept of ANNs briefly in chapter 5. Now we need to understand more specifically how meaningful patterns of information can be captured by complex patterns formed across the massive networks of simple neural connections that make up the mind's ordered anatomy.

The math for representing information in neural circuits was first formulated by McCulloch and Pitts (1943). In one of the famous episodes of modern neuroscience, a distinguished neurophysiologist at the University of Chicago, David McCulloch, took in a homeless teenager, Walter Pitts, from the streets of Chicago. McCulloch recognized that Pitts was a prodigious mathematical talent (Heims 1991). As McCulloch explained the brain's cellular organization and the way neurons sum their synaptic inputs until they reach a firing level (then transmitting across synaptic connections to other neurons), Pitts proceeded to formulate a novel set of nerve net equations to describe the ensemble behavior of simple neurons. The authors of many modern computational simulations of neural networks trace their origins back to McCulloch and Pitts. The goal of neuromorphic neural net simulations is to create input-output relations for the network, similar to what McCulloch and Pitts originally created.

Not long after this work, Donald Hebb, a Canadian physiological psychologist, followed a similar line of reasoning to explain neuronal learning. He recognized that the information gets captured not by the neurons but by the strength and patterns of the connections among them (Hebb 1949). Hebb proposed creative theories that were ridiculed by the biologists and psychologists of the time as unnecessarily speculative. With the hindsight of a half century of progress in computational modeling of brain mechanisms, Hebb's speculations have finally come to be revered by the AI and neuroscience communities.

A model of strengthening single neurons is not particularly useful, given the dynamic interactions of the many connections among the brain's simple and numerous neurons. Hebb proposed that a memory trace could become activated through engaging a *cell assembly*, a group of neurons whose interconnections become stronger with use. The interesting property of this design is that once the interconnections are strengthened, activating any one of them tends to activate, or *ignite*, all of them.

If you think about how a memory gets triggered, often by an apparently random association, you can see, and maybe feel, what Hebb was talking about. The smell of fresh rain on green grass may be enough to bring back a summer from your childhood. In the Beatles song *Yes It Is*, the forlorn young man pleads with his new girlfriend not to wear red on their date tonight because that is the color his former lover wore when she broke up with him. The red of her dress brings back a flood of memories and feelings. The mind's concepts are assemblies of many possible information elements, and when enough interconnections are engaged, the conscious experience may emerge. This is what Hebb called igniting the cell assembly.

The thing that matters is not the cell but the network. The network is defined by the architecture of connectivity that creates a functional pattern within the brain's massive and densely interconnected neuronal populations.

Hebb's realization remains an essential one for understanding both modern AI and human brains. With this, we can look at the patterns of connections in the brain and understand how to interpret its information-carrying capacity from its wiring (connection) diagram (as in figure 7.7).

ANNs became instructive when we could not only speculate about them qualitatively, as with Hebb's theory, but also specify their operation with mathematical models. Captured in math, and soon easily modeled in digital software, scientists could study the properties of specific network architectures.

The function of ANNs in representing patterns of information was advanced with a functioning model in the 1950s at the University of Rochester. In a project for the Office of Naval Research, Frank Rosenblatt created the Perceptron, a machine designed to recognize novel objects from an array of input data. For example, the idea might be to assemble the various signals from a sonar array to identify an enemy submarine. Although he originally implemented the Perceptron in hardware, it was the software simulations that eventually followed that provided the fundamental insight for the next generation of ANN modelers. The input data (such as sonar signal patterns) was connected to simple processing units (dumb neurons). These units were then interconnected with themselves with weights or *strengths* of connection that could be adjusted to represent the patterns of the inputs.

Rosenblatt demonstrated a fundamental principle of information representation in distributed processing units. This principle works for brains as well as for computational models. The principle is that the significant pattern of the information—the thing that reveals the submarine image from the diverse sonar signals—can be captured by the *interconnections* of the dumb neurons that are abstracted from the sensory input units. Fundamentally, the information representations in today's AI systems emerge from this simple, if unintuitive, property. The main difference is that now we add more layers of simulated neurons.

As an example of how a brief and attenuated perception can ignite a larger network of connections, we can return to the example of recognizing your physician in the movie theater. She quickly passes in front of you as she takes her seat a couple rows down. She looks familiar. You only saw her for a second or so. She wasn't looking at you, so you have only this one glimpse, a minimal sample of sense data. Because of your lifelong experience in visual perception, you can extract a lot of information from this glimpse. Even though you weren't expecting to see this person, your brain can still organize an accurate percept, given a little time. What the Perceptron taught us about distributed representation explains how your brain could construct meaning from such minimal data, and it illustrates how neural networks represent (capture) information generally.

Each "neuron" represents only the most elementary visual data, and yet the pattern of the image emerges through the pattern of the many connections among the elemental neurons. The image is then a whole—a gestalt—that is supported by many overlapping and redundant connectional elements. One illustration of the holistic nature of the distributed representation is a feature called *pattern completion*. Part of an image may be missing. Yet, if the neural network has been trained in recognizing that image, the representation (the concept) is delivered up as the whole pattern (albeit slightly degraded). Digital computer programs processing the discrete pixels in isolation cannot do this.

A related property of distributed representations is *graceful degradation:* if there is damage to one part of the network, the network as a whole continues to deliver up the image, more or less. The network is still damaged and the information suffers, but the redundancy and holism of the distributed representation, created by the many overlapping connections among dumb neurons, allows the information function of the network to maintain a kind of global integrity. Contrast this with the operation of your laptop. Information processing proceeds in a linear, sequential logic, and a mistake in one line of code can crash the operating system.

These properties of distributed representation help to explain not only ANNs, but also the pattern recognition that is natural to human perception. Even from a brief glimpse of visual data, the human visual system is able to assemble an entire perception, recruiting extensive patterns of prior experience. In this example of the glimpse of the woman in the theater, you begin with the emotionally significant sense that she is familiar to you as your brain assembles a full, personally meaningful impression from the small sample of initial visual data.

8.1 Deep AI

Operating at a single level of neuronal interconnections, the Perceptron provided us with a first-order intuition for how distributed representations can work to create functional patterns of information. In recent years, the societal impact of AI has mushroomed. Remarkably, the major architectural advance that recent AI models provided over the simple Perceptron architecture has been to add more levels of neural networks to the computational architecture. With enough levels, AI became today's *deep learning*. Understanding the basic principles of deep learning provides additional insight into how we can interpret the brain's connectional architecture as the functional architecture of mind.

Now, you might think that it would have been obvious to the early computer engineers to try adding more levels to Rosenblatt's Perceptron. If so, maybe we would have had usable AI in the 1960s rather than the 2020s. Indeed, the progress in computation in the 1960s and 1970s was with digital computational logic, as engineers recognized the ability to print complex logic circuits as (increasingly complex) solid-state components.

But the difference in AI between digital computing and ANNs may be a reminder of the importance of scientific intuition. Scientists and engineers are trained in sequential logic, so it was straightforward to grind out the increasingly complex digital logic of central processing units. It was not so obvious that the way to create effective AI would be to continue to emulate the processing of neural networks, especially when the first efforts produced only modest pattern recognition. Only a few scientists got this (Grossberg 1980).

A problem with experimenting with the connectional architectures within Perceptron-like computational models was how to adjust the strength of the connections adaptively in order to fit the goal. The goal might be recognizing a specific face, given the training input in the form of images of many varied faces. This became particularly difficult as soon as a second level was added to the neural network. How should the connection weights be modified so that both the input level and the higher (hidden network) level could operate together to improve pattern recognition?

Training the weights is actually a thorny problem. It is the crux of learning. A general approach to converging on a solution in a computational problem can be described as *gradient descent*: if the error between the trial result and the desired outcome is going down, then keep going in that direction. By the 1980s, certain computational modelers discovered how to apply this method to multileveled neural networks. David Rumelhart, Geoffrey Hinton, and Jay McClelland were key figures in the parallel distributed processing (PDP) group at the University of San Diego, studying the properties of various computational architectures for ANNs (Rumelhart & McClelland 1986). They showed that tracking the improvement of errors could be transferred across two or more levels of network architecture in a method that came to be called *back-propagation*.

The method was somewhat artificial, and at first it didn't seem so brain-like: after computing the error between the network's representation and the desired goal, you poke that error back into each level so that the network adjusts to minimize the error. Although it was arbitrary and artificial, once back-propagation was applied to allow error minimization, the pattern learning capabilities of these multileveled networks proved to be quite brain-like, providing a new way of reasoning about how the actual neural networks of real brains might achieve their pattern recognition (conceptual) skills (McClelland, McNaughton, & O'Reilly 1995).

By the early 1990s, the advances in PDP modeling of ANNs had progressed to the point that it seemed—at least to some theoretical neuroscientists—that a new synthesis was at hand, with the computational modeling providing a strong basis for understanding how arrays of neurons represent information. These computational insights promised ways of understanding the new neuroimaging methods of positron emission tomography (PET) and functional magnetic resonance imaging (fMRI), as they showed reflections of the actual neural functions of people's brains (Changeux & Dehaene 1989; Posner & Raichle 1994; Tucker 1991).

Neuroimaging research became immediately popular and advanced rapidly. However, the computational insights from neural networks did not seem to be relevant to the neuroimaging researchers of the 1990s and 2000s. They were content to catalog the functional correlations of brain regions with experimental tests of cognition and perception. The advances that proved necessary to spark the recent progress in AI did not come from the thinking of scientists of this era. Rather, the necessary advances to generate modern AI appeared from two developments in modern culture that required another couple of decades to appear. We could say that modern AI came not from theoretical insight but from cat videos and video games.

To be precise, it wasn't just cat videos. Rather, it was the massive quantity of images of all kinds posted on various internet websites. Within a couple of decades, researchers organized photos with search engine technology to allow access to large numbers of examples for tasks such as AI for image recognition. The result was the creation of what would come to be called *Big Data*. Although most statisticians understood the power of large numbers, it was a striking realization that AI algorithms for tasks such as image recognition would give exceedingly weak performance when trained on 100 images, but they yielded highly accurate performance when trained on 10,000 images.

The second advance came from an unexpected arms race that emerged in marketing home computers. To meet the demand for fast graphics in video games, several companies developed graphics processing units (GPUs) to accelerate the many simple computational operations required to move avatar heroes, monsters, and explosions rapidly around the video screen. Because these GPUs were essentially parallel computers, once they became available, engineers took them as inexpensive supercomputers. One of the obvious applications was multilevel ANNs, which required big computer power to process the Big Data of internet uploads of the artifacts of modern culture (party pics, real estate ads, selfies, and cat videos). By the 2010s, after a couple of decades of disinterest in ANNs, researchers demonstrated impressively accurate pattern recognition with multilevel neural architectures that were extended up to ninety levels.

There are two clear insights to be gained from the recent successes of AI. The first is what is called "the unreasonable effectiveness of data" after Wigner's classic quip on the unreasonable effectiveness of mathematics. Big Data allows solutions that generalize in a solid way that is not possible with algorithms based on small samples. As just mentioned, recognizing cats in internet videos was not possible with 100 videos, but became highly accurate when the internet provided 10,000 cat videos. The second insight is that of deep learning: organizing the networks of neurons into multiple levels (not unlike those in figure 7.7) allows specialization of each higher (deeper) level for more abstract representations. Whereas the connection patterns for any given level are structured for the specific representation at that level, a higher level can form new connection patterns that are then linked to the lower level but not constrained to the pattern at that level.

Now, creating AI with ninety levels may be effective for recognizing cats or detecting fraudulent patterns of credit card use, but this is of course beyond the four major levels (primary, secondary, heteromodal, and limbic) of the human cortex. In this way, modern AI is forging ahead as an engineering discipline with its own novel architectures implemented in increasingly sophisticated electronics. Once we appreciate what is actually required to form concepts, we might worry that AI will soon leave the human brain behind.

8.2 Levels of Human Knowing

Through intuitions developed from simulating large-scale neural networks, we gain new insights into how the embodied human mind emerges from the architecture of the brain. The computational insights into multileveled distributed representations converge with the anatomical insights into the actual connections of human corticolimbic networks (such as the four levels or types of cortical structure described in chapter 7). This convergence teaches us how to theorize more clearly about the multileveled process of knowing.

In vision, for example, the primary visual cortex extracts relatively primitive features of the retinal map, such as receptive fields organized in relation to lines and contours. This primitive abstraction (summarizing the raw data in more compact form) allows the next level—the secondary (unimodal) visual cortex—to extract even more abstract concepts of objects (in the ventral "what" pathway) and their movements in space (in the dorsal "where" pathway). In the heteromodal association cortex there is further abstraction of concepts not just of one sense but of the multisensory convergence of experience. In chapters 10 and 11 we will explain how we understand the term "concept," not as an abstract mental entity but as a pattern of neural connectivity that consists of expectancies (as affordances) for possible experiences.

As we saw in chapter 7, the highest level of processing—above the heteromodal cortex—is achieved in the core limbic networks, which are responsible for mediating cognition with internal bodily states. Whereas primary sensory and motor cortices are modular, specialized for one sense or the actions of one side of the body, the association cortices (unimodal and heteromodal) are more general, integrating multiple forms of information. Judging from their denser connectivity, the limbic cortices have evolved as the *most* generalized convergence zones. The straightforward implication is that the conceptual representations in limbic cortex are the most generalized and holistic within the brain's architecture.

In these limbic networks we expect to find what the neurologist Antonio Damasio (1999) called "the feeling of what happens." This is the feeling of your body states being changed by your ongoing interactions with your environment. In one sense, this may be the most abstract of the brain's representations, covering the large space of meaning that includes specific association cortices. The interesting paradox is that the limbic part concepts must be concrete as well, constrained as they are by the primitive subcortical circuits of motivational control (Derryberry & Tucker 2006). True abstraction may emerge only through the integration across large-scale cortical networks at multiple levels of representation (Tononi 2008).

Interpreting the human cortical network architecture with principles of distributed representation suggests how information is organized in the mind. In a naturalistic, scientific philosophy we accept that concepts are isomorphic (of the same form) with our cerebral networks, such that the process of knowing must be organized in this exact neural architecture. We make contact with the world through primary sensory cortices on the one hand and through primary motor cortex on the other. Although they do not stand out in experience, we can infer that the unimodal and heteromodal association cortices each provide levels of conceptual integration within the process of conscious experience. Then, at the core of the hemisphere, in the limbic cortex, is the highest level of integration.

Let us pause for a moment and take stock of the important implication of the process of experience that we have just described. *From our description of the four levels of organization, the most general level involves an affect-rich holistic sense of our present experience as full of meaning.* This is the point that we took Dewey to be making when he said that objects and events stand out and have meaning only in relation to a qualitatively rich background sense of a unified situation. This holistic limbic representation comes from our deepest visceral homeostatic processes that are crucial for our survival and well-being.

How do we understand this architecture as the basis for knowing? Can we understand the process of cognition as a negotiation between an integrative visceral base of experience—the feeling of what happens—and a somatic differentiation of the specific form of the contact with the world? We suggest that we can. Science is charting the blueprint for the phenomenology of experience.

8.3   Cognitive Control at the Limbic Core

As neuroimaging technologies have developed over the last several decades, including PET, fMRI, and dense array electroencephalogram, the assignment of specific cognitive functions to specific cortical regions has been supported by powerful new evidence. This evidence has provided functional demonstrations that the human brain's cognitive capacities require integration across large-scale networks. This research illustrates another important direction of scientific progress that leads to an embodied, naturalistic philosophy of mind.

A cognitive task, such as reading a map, engages visual and visual association areas and also frontal networks that are involved with action planning and execution. It is as if the interpretation—even though it seems perceptual—requires some covert action planning.

Perhaps more surprising in this neuroimaging research was the regular engagement of limbic cortex when difficult decisions had to be made. For example, imagine a simple task that challenges your executive control. This task requires you to press the right button quickly when an arrow points to the right, and the left button when the arrow points to the left. Then, on some trials there are two arrows, pointing right *and* left, and you must not press *either* button. This simple ambiguity of conflicting arrows causes more errors, and requires you to monitor your actions more closely. This greater monitoring to make a correct response engages the anterior (frontal) regions of the limbic cortex in the cingulate gyrus (Luu & Tucker 2001; Luu, Tucker, Derryberry, Reed, & Poulsen 2003). The implication is that the ongoing motivational control from limbic regions is essential to the frontal lobe's executive control of action.

Clarifying the role of limbic (visceral) networks in regulating human cognition may be one of the most interesting theoretical challenges in cognitive neuroscience (Barrett 2017). Many neuroimaging studies show that as people perform cognitive tasks, particularly if effort is required, there is often strong activity in the limbic regions of the brain. These regions contribute to what has been called the *salience network* (Seeley et al. 2007) insofar as they determine how cognition is shaped by values concerning what is important to continued functioning and well-being. Furthermore, in terms of architecture, all of the association areas of the cortex—the generalpurpose networks—are interposed between limbic cortex and either the sensory or motor regions of the neocortex.

As we have seen, the visceral component of the limbic system regulates the body's internal state in order to preserve, restore, or enhance bodily well-being. It also provides motive control of memory consolidation for the entire cerebral hemisphere (Mesulam 1990). To summarize: *As the somatic sensory and motor networks of the neocortex regulate the traffic with the external world, the visceral networks of the limbic cortex regulate the brain's traffic with the internal milieu of bodily needs. These needs mediate between internal homeostasis (and allostasis) and the motivational control of external perception and behavior.* What has been a striking surprise over the last half century of neuroscience research is the discovery that the control from limbic regions is also essential for regulating the process of memory consolidation throughout the cerebral hemisphere. This convergence of motive and memory functions is perhaps the most important key to understanding the mind's activity within the brain's architecture.

Consistent with a central role of limbic regions in controlling memory, neuroimaging studies have shown that there is often activity in the limbic, salience, networks, even when the person isn't obviously emotional—or motivated—but simply working to perform the cognitive task.

We might think that the distinctively human frontal lobe of the brain would be the unique active network for executive self-control, that is, our ability to monitor our cognitive activities of thought and action. Yet, the evidence proved *not* to line up with this preconception. Instead, the most complex and sophisticated forms of control over cognition turn out to engage the limbic cortex—such as the anterior cingulate—as well as neocortical frontal networks. For example, the early neuroimaging studies (using PET or fMRI) showed that the anterior cingulate cortex engages whenever cognitive control is required, even for complex decisions (Posner & Rothbart 2009). *It became increasingly clear that what we had thought of as purely cognitive acts are profoundly shaped by our deepest visceral needs and values.*

Another example comes from neuroimaging studies showing strong limbic engagement when subjects engage in self-relevant cognition (Pfeifer & Allen 2012; Tucker et al. 2003). An example might be judging whether traits, such as *hardy* or *flexible*, apply to you. The evidence shows that cognition relevant to the self seems to require the involvement of the limbic cortex and the association cortex, such as the medial frontal lobe, which elaborate limbic representations. Your identity is deeply rooted in your visceral needs and drives.

How could these complex functions, such as self-evaluation or executive control, emerge from the visceral regions of the brain? In the early days of cognitive neuroscience, it was hard to imagine a cognitive scientist admitting that the brain's visceral networks could be required to exercise the mind's most complex executive control of its own operations.

Yet, this is what the evidence revealed. This involvement of the visceral networks in high-level cognitive operations represents the embodiment of mind—both at the visceral core and at the somatic interface with the world. Accepting this embodiment of cognition in the brain's physical networks is necessary to explain the data of the experiments. This is not only because we observe activity in the limbic networks during cognition, but also because there isn't anywhere else for it to happen. There are no purely cognitive modules that are separate from the bodily (visceral and somatic) functions.

The important conclusion we draw from this body of research is that all of our cognition is shaped by limbic structures responsible for discerning salience and for visceral monitoring of homeostatic and allostatic processes, geared toward the well-being of the organism. There is no disembodied, value-neutral, or emotion-free cognition. Cognition evolved to serve action, and action must be directed by motivational systems to provide value to the organism.

8.4   The Unconscious Consolidation of Experience

Even a superficial analysis of the brain's network architecture is incomplete if we cannot say how information is represented in that structure. We suggest that neural network AI provides essential conceptual tools for interpreting the neuroimaging data. The structure of neural networks is defined by patterns of connections.

The task of growing the brain's connections throughout life is *learning*; forming new synapses (connections between nerve cells) and modifying old ones to capture the meaning of what happens. This capturing of information within patterns of neural connections is integral to the process of experience. The brain's synaptic architecture directly reflects the structure of mind. The neurophysiological process through which this happens is called *memory consolidation*. New learning is maintained through the activity of limbic circuits, which then interact across the heteromodal, unimodal, and primary cortical levels to record what has been learned in the synaptic architecture of the cortical networks.

The necessary operation of the limbic system in memory consolidation helps to explain why the large-scale architecture of the forebrain is organized around the limbic hubs of the hemispheres: these are the adaptive controllers of learning and memory, regulating the motivational basis of knowing.

The central role of limbic networks is clear from the classical evidence on memory consolidation. It was long known from clinical neuropsychology that damage to the temporal lobes and their limbic cortex may result in severe memory loss, called *amnesia*. This was confirmed by the severe anterograde amnesia (inability to remember new things) caused in a famous case (H. M.). In an effort to control seizures, H. M. underwent bilateral surgical removal of the limbic structures of the temporal lobes. This man then seemed relatively normal in his cognitive function, except that his world had stopped changing. He remembered nothing new from the time the surgery was performed. The implication was unavoidable: the limbic structures and their associated limbic cortex play an essential role in organizing memory.

Animal research and additional human clinical observations clarified this process of memory consolidation further (Squire & Zola 1997). Contrary to our commonsense assumptions, memories are not formed instantaneously. Rather, they require a neurophysiological process over time in which to organize the memory traces enduring in cortical networks. If a person suffers a seizure before the consolidation process has proceeded sufficiently, the memory for that episode is lost. If the seizure occurs after consolidation has been sufficient, the memory is retained.

In his training as a clinical neuropsychologist, one of us (Tucker) accompanied a depressed patient to her electroconvulsive therapy (ECT) sessions each morning for a week. Even though they talked at length each morning, the patient would greet him as a stranger each day because her memory of the previous day's meeting had been erased by the ECT seizure.

The limbic system regulates how we capture new learning and how we transfer it to the neocortex through the interconnections shown in figure 7.7. Once transfer is complete, the limbic structures are less important, apparently because the neocortex is now the primary site of the memory storage.

Although some transfer of volatile memory to a more stable form in the neocortex is required for memory consolidation, the stability of memories is relative. Considerable evidence shows that we continue to consolidate memories not only for hours and days, but for months and years (Squire &

Spanis 1984). In one classic study, depressed patients were examined who had a series of electroconvulsive treatments, each one causing a seizure that disrupts memory consolidation (Squire, Wetzel, & Slater 1978). The effects of the seizures on the retention of long-term memory were calibrated over many past years by testing for memory of events in popular television shows that had gone off the air some years before. The patients' memory loss showed a consistent temporal gradient: the more extensive the series of electroshock treatments, the longer into the past the memory loss extended (Squire et al. 1978).

The remarkable implication is that our memories of past events are active in the unconscious mind for many years. We may seldom think about old television shows. Yet, we seem to dream about them, or otherwise keep their memories engaged unconsciously. The vanishing temporal gradient of memory loss following electroconvulsive seizures implies that some neuro-physiological process is continuously active within the unconscious mind to maintain these past memories.

Other evidence suggests that, once reactivated, memories are then more susceptible to interference or loss (Squire et al. 1990). Accessing conscious memory somehow changes the nature of the background neural activity from a less volatile (and less accessible) to a more volatile (and more accessible) form. There thus seem to be degrees of unconsciousness.

The evidence on memory consolidation provides fascinating clues to the ongoing and dynamic activity in the unconscious mind. A moment's reflection makes it clear that a functioning memory is essential to the ability of our minds to bring personal experience to bear on current thinking. It is remarkable that personal experience—the dynamic contents of the self—is embodied in ongoing neural activity, which continuously shapes the connections among massive neuronal networks. Through recognizing the limbic base of memory consolidation, we can understand the central position of the limbic core in integrating the cognition of the cerebral hemisphere. We can also appreciate the role of the visceral evaluation of personal significance in organizing cognition in relation to the somatic interface with the world. The ongoing process of memory consolidation appears to ripple up and down the cortical networks—from the interior visceral limbic core to the exterior somatic contact with the world and back again. This seems necessary, not only over the first minutes required for initial memory consolidation, but also over the months and years required to integrate personal experience within the enduring memories that comprise the self.

In this process, there is a special role for the unique neurophysiology of memory in sleep. Increasing evidence shows that each day's experience must be integrated within the neuronal ensemble of the organism—the personality—in each night's sleep. This integration engages the full corticolimbic architecture that we have studied in this chapter.

Whereas memory consolidation is ongoing to some extent in the background of the waking mind, this process seems to receive unique and important contributions from each stage of sleep (Walker 2008). If sleep is disrupted to a certain extent, then memory is impaired to the same extent. In the deep stages of sleep, the neocortex is synchronized by large slow electrical waves. This period of *slow-wave sleep* has been found to be essential for organizing both motor skills and declarative memory, such as remembering items on a test. In contrast, the rapid eye movement or dream sleep, which becomes more frequent toward the morning, seems to allow recent emotional events to be integrated in some way with past experiences (Cartwright, Agargun, Kirkby, & Friedman 2006).

8.5   The Motivated, Value-Laden Roots of Cognition

These findings provide a picture of the limbic consolidation of memory, the ongoing process of organizing experience, and the adaptive control of the mind's capacity. The major insight gained from considering the large-scale architecture, the essential role of limbic regions for consolidating memory, and the tight links between limbic networks and the homeostatic and allostatic controls of the visceral function is that *learning and memory—and the mind's essential foundations—are regulated by adaptive, emotional, and motivational controls.*

These are active controls, such that the functioning mind requires continuous neurophysiologic activity of a sufficient quality if it is to retain the continuity of the self. The devastating effects of dementia illustrate the loss of mind when the neurophysiology of memory degrades. At first, memory loss affects recent learning primarily, such that the person's historical self is relatively intact. However, progressive dementia causes the history that defines personality itself to dissipate, such that only a shell of the self remains.

Operating through the network architecture of the limbic, association, and primary sensory and motor cortices, the neurophysiology of memory consolidation maintains the growth and development of the brain. It changes the synaptic connections among neural networks, such that the ongoing function of mind is continually aligned with the architecture of the cortico-limbic networks. As we build the science of human memory, we gain fresh perspectives on the anatomy and physiology of the unconscious mind that continuously refreshes the developing self.

The unconscious integration of daily experience with past memories during sleep implies that memory is not a simple copying of the traces of perception or action into storage, but rather an organizational process in which the ongoing neurophysiology of brain function—the implicit self—is modified by significant events in the process of experience. When the events of the day have strong emotional significance, as when children encounter a scary experience, the dreams of the night become intense, as if the integration within the ongoing neurophysiology of the self is both strongly engaged and difficult to achieve. In a similar way, the distressed and repetitive dreams of posttraumatic stress disorder suggest that a failure to achieve the reorganization of the self may lead to continuing adaptive failure (Cartwright et al. 2006).

8.6 The Anatomy of Subjective Experience

Recognizing that the mind arises directly from the brain's interaction with our body and its environment, we can realize how a person's conceptual system is organized through the architecture of the cortical networks. Although we know that mental processes are distributed across multiple linked networks, we can see how concrete images must arise through links to the sensory cortex or links to action representations in the motor cortex. In contrast, the feeling of significance must arise through links to the limbic cortex. This is the feeling of what happens (Damasio 1999). It is a somewhat vague and implicit quality of consciousness, but essential to understanding the significance of events.

It may be within this interchange between the feeling of significance and the perception of ongoing events that the continuity of experience arises. Consciousness requires this continuity of experience in time, what we call *ongoing memory*. Gerald Edelman emphasizes the essential role of ongoing memory in consciousness when he describes "the remembered present" (Edelman 1989; Edelman & Tononi 2000).

The clinical evidence shows that if ongoing memory is disrupted, such as through a head injury or seizure that interrupts the limbic consolidation of memory, then not only does the person lose consciousness immediately, but they lose the immediate memory for where they are and what's happening. They wake up disoriented and confused. With a more extensive injury or prolonged seizure, amnesia extends longer into the immediate past. A poignant example of these types of functional breakdown is the popular noir film *Memento* (2000).

This evidence implies that we are always consolidating the recent past as the context for present awareness. This context provides the continuity of the implicit self. It is within the resonant processes in the corticolimbic architecture, continually organizing memory, that the subjective experience of self—and consciousness—arises.

Understanding the motive control of the memory process provides an interesting perspective on the nature of consciousness and its emergence from unconsciousness. It is *not* like Freud said: that the unconscious is a disavowed domain of the mind, with its threatening contents kept from awareness. Rather, the visceral motives of the unconscious mind must be the wellsprings of consciousness, resonating with the ongoing processes of perception and action to select those elements of memory consolidation that are sufficiently sustained to become conscious.

To imagine how experience arises from the linked networks of cortical representation, return to the example of figuring out why that woman you saw in the movie theater looked so familiar. The network architecture helps to explain why the process of perception involves not only the sense data in primary visual cortex, but also the memory of previous perceptions that serves to organize the sense data. As limbic and association networks participate in the perceptual process, through the ordered network connections shown in figure 7.7, each level brings the abstract representations of prior experience to constrain the input data from each lower level.

In reflecting on the glimpse of the woman in the example above, the multileveled nature of your perception is essential to the psychological experience. You organize the visual data from the retina and the thalamus in not only the elementary lines and contours, but also the configurational face patterns gained from experiences in human social interaction. How you evaluate the face pattern as familiar almost certainly depends on the limbic cortex as well as the association cortex. The sense of familiarity is described by psychologists as a *feeling of knowing* (Shimamura & Squire 1986). It is indeed a feeling, with the inherent motivational significance that causes us to value the process of knowing intrinsically. It is not practically significant that you identify every person you see while in the theater. Yet, the feeling of familiarity in this glimpse becomes a motivator for exerting mental effort to extract the appropriate memory associated with this perception.

In this example, the process seems less like the instant ignition of your Hebbian cell assembly and more like a recalcitrant smoldering—you sense there's some heat but you can't see the light. The perception began as a kind of visual recognition, a diffuse feeling of familiarity that was engendered by seeing this woman. We can infer that the input pattern in the visual cortex must have resonated to patterns in higher-level (heteromodal association) networks, which in turn gained some sense of significance from your visceral limbic response to the brief glimpse. You may notice a vague feeling of personal significance associated with this glimpse. Recognizing and reflecting on this, you might finally remember that this is your doctor, in jeans rather than a white coat.

The cognitive process starts with a diffuse sense of meaning and then over time (seconds and milliseconds) organizes a more articulate and differentiated representation in mind (J. W. Brown 1994; Kragh & Smith 1970; Tucker 2007). The levels of meaning in this process depend on the organization of the corticolimbic neural architecture that we have studied. The visceral base provides the integrative, and generative, core. Linked to the motive control from this core, the extensive association cortices then engage the more articulated sensory and motor representations in primary somatic cortices to allow distributed patterns of conscious conceptualization to span between the visceral and somatic boundaries of mind.

Several lines of evidence converge to clarify the architecture of the embodied mind. Perhaps most important are the advances in connectional neuroanatomy, so that we can map the cortical networks in detail. As a result, we can study precisely the ways they must represent the structure of experience. Through computational models of ANNs, we gain intuitions for the nature of distributed representations, and these then allow us to appreciate how function arises from the specific connectional architecture. Through understanding the way that cognitive activity engages the multiple networks, and the essential traffic between limbic and neocortical networks that allows memory consolidation, we begin to appreciate the neural mechanisms of mind, not in some experimental paradigm but in the integration of experience in each day's events—and each night's dreams.

8.7 Shifting Attention between Visceral and Somatic Constraints

Thus, the mind's embodied elements—its concepts—can be understood not only in relation to the *somatic interface*, but also in the feelings and motive urges that signal the personal biological significance, the *visceral base* of the self. The clear implication of the anatomy of memory is that concepts can only be organized (and then consolidated in sleep) by combining *both* visceral significance *and* the concrete somatic interface with the environment, as the reality of the local context is perceived and manipulated. There is nowhere else that concepts could arise. Recognizing that both domains—visceral and somatic—are essential, we might wonder whether these visceral and somatic constraints regulate the mind's operations to different degrees on different occasions?

Sigmund Freud would certainly have answered yes. Impressed with the discovery of the mind's evolution, Freud viewed the mind's most basic operation as driven by the visceral needs—the personal biological motives—manifested impulsively in dreams and fantasy in *primary-process* cognition (Freud 1940). In contrast, meeting the demands of the world, both in actions and in reality-tested cognition, he described as *secondary-process* cognition. It is through the somatic sensory and motor networks that we make contact with the reality of the world.

Findings from modern neuroimaging experiments may be consistent with Freud's distinctions. When researchers examined the patterns of correlation among the activity levels of cortical regions with fMRI measures, they found that certain regions varied together in time (Obrig et al. 2000). These regions somehow worked together, forming a functional *network*. We have already described two of these basic corticolimbic systems, one for the concepts of the general configuration of the spatial context (dorsal) and another for the specific objects that have significance in that context (ventral).

Although these large-scale network patterns fit with both human and animal evidence on attention and memory systems, researchers were surprised to find networks activated when subjects were *not* paying attention to a particular task in an experiment. Certain limbic regions, such as the posterior cingulate cortex and the medial frontal cortex, together with certain related neocortical association areas, became actively integrated in those rest periods when subjects were *not* required to perform the task of the neuroimaging experiment. These were times when the subjects rested quietly. This brain pattern was then described as the *resting-state* network. Because subjects seemed to return to this network when not specifically instructed to pay attention, it was also called the *default-mode* network (Raichle & Gusnard 2005).

This pattern was surprising. Some researchers asked the subjects what they were thinking about when the default-mode pattern appeared. The subjects replied they were thinking about what they were going to have for dinner, or the shopping they needed to do, or what they had planned for the weekend. This undirected mental activity, related not to the tasks of the experiment but to salient personal values, was then described as the *mind-wandering* network.

Thinking back to Freud's ideas of primary and secondary cognitive functions, and recognizing the strong limbic involvement in the mind-wandering network, we could also consider this as the *primary-process* network, forming concepts that are more attractive to personal interests (and fantasies) than to interfacing with the current events of reality.

Considering the overall architecture of the cerebral hemispheres that is required for memory consolidation, we speculate that these network patterns reflect the dual constraints of the telencephalic hemispheric architecture. The somatic constraint links the mind's operation to the body's interface with the world when cognitive tasks are required—in either the dorsal or ventral attentional mode. The visceral constraint—set loose in the default mode of fantasy and mind wandering—is responsive to the limbic control of personally significant concerns.

There is a familiar phenomenology of experience suggested by this architecture. Our experience of sensing aspects of our surroundings and acting in those surroundings is Freud's reality-testing or secondary-process cognition. The differentiation of experience, in articulating the features of sensation or the discrete movements of actions, occurs at the somatic interface with the world in the sensory and motor neocortices. In contrast, judging by the hemispheric architecture as shown in figure 7.7, the integration of experience, where the multiple elements come together, must be at the visceral core of the hemisphere. At the core, emotional resonance and self-relevance provide motive controls that maintain significant activity over time, thereby consolidating memory. When engaged as the primary direction of the mind's operations, the visceral core tends toward primary process cognition, where the constraint doesn't fit the reality of the environment, but rather resonates to the needs and wishes of the self.

In the shifts between attention state networks and default mode network, we see the mind alternating between somatic and visceral constraints. Even though alternation is often observed, it is the simultaneous negotiation of these constraints that is necessary for the enduring organization of experience in memory, whether in waking reflection or in each night's sleep.

8.8   The Adaptive Unconscious: A Summary

Let us summarize our argument in this chapter. We have briefly reviewed the anatomy of the human brain based on the realization that the mind's structure is revealed by the brain's connectional architecture, which is the product of evolutionary development and subsequent interactions with our physical, interpersonal, and cultural environments. We saw how principles of distributed representations in AI systems provide insight into the functional architecture of the brain—its vast patterns of neural connectivity that make possible cognition and action. In order to be linked to the world, concepts must engage the sensory and motor representations of the primary cortices. In order to be significant for the person, concepts must be grounded in the limbic networks of feelings and motives. Between somatic and visceral constraints, the association networks provide not separate cognitive modules but further distributed network links. As a result, the cognitive process—the process of knowing—emerges through the functional negotiations of these multiple levels of the visceral-somatic hierarchy.

Knowing requires the consolidation of memories, which establish the context and the concepts for our learning. The evidence on memory consolidation shows that whereas memories may be represented functionally in the neocortex (association and sensory-motor cortices), they can only be consolidated—organized for retention within the neural structure of the self—through the action of the limbic cortex. The clear implication is that cognition depends in each moment on the adaptive control arising from emotional and motivational concerns. If we take the evidence of anatomy and physiology seriously, then the concept of the cognitive unconscious must be expanded to encompass the adaptive unconscious, the motivational and emotional influences that continually shape the process of memory in the ongoing background of the mind during each moment of waking and sleeping.

Reflecting on the biological evidence thus leads to an important discovery for the process of knowing. Cognition cannot be separated from emotion. If thoughts are to be retained long enough to become organized within the mind, they must be consolidated by the memory mechanisms, which are regulated by the limbic, emotional and motivational, networks. If we align the structure of concepts literally with the structure of the corticolimbic architecture, then both the limbic adaptive base and the sensory-motor articulation must be integral to each concept.

We can now see that this combination of limbic and sensory-motor shaping of concepts applies not only to concrete or perceptual concepts, but also equally to abstract concepts, which are defined mostly by metaphors. We saw how the source domains of the metaphors, which are appropriated for structuring of the target domain, have their meaning precisely via the sensory-motor, motivational, and emotional dimensions of our mundane experience. Embodied meanings are recruited for understanding more abstract concepts (such as mind, thought, knowing, freedom, will, goodness, etc.). Recall how our metaphors for thinking (e.g., THINKING IS PERCEIVING, THINKING IS MOVING, THINKING IS OBJECT MANIPULATION, THINKING IS PREPARING AND EATING FOOD) all involve sensory, motor, and affective dimensions of bodily experience whose spatial or corporeal logics (or inference patterns) are recruited for metaphorically structured conceptualization and reasoning.

Of course, these neural foundations of mind must be inferred from the scientific evidence; they are not readily apparent in conscious introspection. We can agree with Hume that when we reflect on experience, we find the usual scattered array of images, feelings, and impressions dancing in the theater of consciousness—without any palpable self behind the curtain to direct the play. Yet, as we learn of the evidence implying the mind's architecture, rooted in the limbic core, we can make the strong inference that the unconscious foundation of experience—the self—is indeed present but implicit. It is continually tuned by the adaptive concerns of the organism, the motives of the self. Each morning as we awaken, we reassemble the implicit self as the organizing context for experience from the residuals of the night's consolidation process. These residuals reflect more than the homeostatic needs of the immediate biological organism; they result from the lifetime developmental consolidation of the embodied mind.

Mind is thus grounded in the elementary functions of the body—feelings at the limbic core and physical contact with the world at the sensory and motor interfaces. This is a different perspective than the linguistic, symbolic account of mind that has come from analytic philosophy and first-generation cognitive science. Human thought is indeed structured powerfully by language, allowing the sharing of meaning through the verbal semantics of the culture, and the order of reasoning through the grammar that is shared by the citizens of the culture. But the neuroscience evidence shows how even linguistic concepts must be organized from the neural substrate. Hence, all of this meaning is profoundly embodied and visceral. Furthermore, the scientific insights from the intelligence achieved by ANNs teach us how this may occur. The model of the embodied mind that has come from studies on metaphors turns out to be highly informative for the scientific study of mind. Human cognition, in language or in any form, is organized from the templates of experience that allow us to expect the regularities of future events. Metaphors of mind thus provide a very general model as we learn more specifically how mind could be embodied within its neural architecture.

In the next chapter we will look further into the adaptive control on the mind's biological process, considering not only the implicit consolidation of experience but also the active, motivational control of learning as the core mechanism of neural development.

9 The Motive Control of Experience

Chapters 7 and 8 set the stage with the mind's architecture outlined in both *structure*—the brain's connectional anatomy—and *function*—the principles of distributed neural representation that teach us how the anatomy works. In this chapter we bring into our narrative of cognitive development a new and vitally important character, namely, the motivational systems that shape all cognition and action. *The remarkable conclusion is that there is no sharp line that separates cognition from its adaptive emotional and motivational base*. All cognition is motivated and influenced by our deepest values tied to our well-being. The assumption of pristine, value-neutral objectivity that is so pervasive in Western philosophy and science must be given up in favor of a more biologically based conception of knowing as shaped by our deepest values arising from our ongoing biological efforts to achieve well-being.

9.1 Motive Control of Neural Development

The adaptive control of thought arises through the self-organizing process of neural development. It is this process that gives rise to, and constantly tunes up, the patterns of neural connections that we examined in the previous chapter. The adaptive control of neural connectivity—organizing neural networks to meet our needs and urges—generates the informational complexity of the mature mind.

The neurophysiological question is how to understand the neural control systems that regulate the brain's development over time, creating a unique mind in the course of development. We examine how the control of neural activity allows enduring patterns of meaning—concepts—that come to be valued and regulated to be adaptive for the tasks of life. We begin this chapter with the basic mechanisms that control arousal in the brain's developing neural networks, thereby biasing the processing of information at the deepest levels of motivational control. In short, these basic arousal control systems create the most fundamental values that shape our experience and knowledge.

We then shift the scope of the discussion to consider the cortical networks that must be regulated by these adaptive motive controls. From research on animal learning, it has become clear that mammals learn through a *cognitive* process, a process of acquiring knowledge. In mammals, the motive controls on behavior operate not only to restore *homeostasis*, responding to immediate needs and drives, but also in an *allostatic* manner as the animal anticipates both its needs and what it expects to encounter in its environmental context. It accomplishes this predictive task by drawing on its mammalian memory capacity in order to construct a primitive expectancy for what happens next. The *feeling of what happens* is an important base for experience, but to be fully adaptive it must generate the feeling of what happens *next*.

This predictive process proceeds first through actively organizing expectancies of what the world should be like and then registering whether these expectancies fit the evidence of the world—evidence that we contact through the senses. In this way our minds continually maintain a context model, a feeling of what's happening, while also reaching into the unfolding future. If the context model collapses, due to an event that's unexpected, we respond with anxiety and a fight-flight response to adapt to the errors of our ways.

This predictive process, Dewey's need-search-satisfaction, is at the heart of our personal acts of knowing. From the biological perspective, knowing can be understood to emerge from the active motive control of the neurodevelopmental process. In this context, we need to consider two principles of neural development that explain the adaptive control of thought throughout life.

The first principle is that concepts, the representational and enactive elements of thought, are formed through the patterns of connections among neurons. As a result, thought is a biological process. It can be analyzed directly in relation to the formation and elimination of neural connections as the mind/brain develops.

The second principle is that neural connections are formed and maintained through the control of neural arousal. Connections are maintained only if they are active. This means that the mind's growth is determined by the control of neural activity in time. There are elementary mechanisms of activity control (arousal) that seem to have been elaborated through several waves of increasing complexity in the evolution of the vertebrate brain. To illustrate these with a simple theory, we will emphasize the elementary mechanisms of *sensitization*—increasing neural responsiveness over time—and *habituation*—decreasing neural responsiveness over time.

These primitive motive controls are integral to the differing responses that occur when events are consistent with the context model (the feeling of what happens) versus discrepant with it (generating fight-flight). Habituation occurs when events fit the context model, and our experiences consistently meet our expectancies for what will happen. Because there is little information we soon lose interest. In contrast, when discrepancies occur, the primitive neural response is sensitization, supporting the defensive fight-flight response. In the cognitive domain, sensitization leads to heightened arousal and a focused attention on threat objects.

The key point here is that we can understand and explain the cognitive effects of these primitive controls on neural activity over time. In maintaining the context model, the dorsal limbic circuits regulate the dorsal brain through a habituation bias that, by refreshing working memory quickly, expands the scope of mental representation for the broader context. In contrast, the ventral limbic circuits tune cognition with a sensitization bias, leading to greater constancy and increased focus of the cognitive process that is required under threat.

The elementary control of neural arousal thus applies dual biases, each favoring a different trade-off of spatial and temporal information. In this chapter we will trace how these elementary arousal controls have been elaborated in mammalian brain evolution to organize memory consolidation in different ways, thereby shaping the active, motivated control of cognition in time. These unique motive controls are different ways of tuning the structure of concepts.

Even for the complex human brain, it is important to understand the continuity of development with the basic vertebrate plan, so that we appreciate the unique adaptations as human brains grow for many years within the rich context of their cultures. At the beginning of life, the neural tissues differentiate from their primordial stem cells and organize the structures of the vertical hierarchy (figures 7.2 and 7.3). Through differentiation, the embryonic stem cells take the multiple specific forms required for the maturing organism's anatomy. The neuronal stem cells of the embryonic brain migrate to their specific positions (following genetically determined chemi-cal gradients), thereby differentiating into the specific neural elements of the six layers of the neocortex.

Yet, these genetic plans for tissue differentiation provide only general guides to creating a functioning brain. At each phase, the architecture of the neural networks must be corrected and refined by a process of neural epigenesis, thereby achieving neurodevelopmental self-organization. The activity of the neurons, tuned by sensitization and habituation, is what determines which connections are useful and will be retained. The unused connections, and eventually the ineffective neurons, wither and are reab-sorbed into the body. The result is preserved patterns of neural connectivity across functional brain regions that become reinforced by the recurrence of certain experiences.

Because human development begins with such an extended juvenile period (now two decades and counting), human genetics provides only a rough out-line. The epigenetics must self-regulate effectively over an extended interval, or else there will be serious developmental pathologies. These include the mental disorders that, compared to simpler mammals, seem to be the all-too-frequent consequence of human developmental complexity.

Motive Control of Habituation and Sensitization in Organizing Experience

This neurodevelopmental process in the embryo is guided from the outset by the arousal control systems of the brain stem, which mature early and set the tone for the activity of specific neurons that allows them to participate effectively, or not, in the neuroembryonic developmental process. For exam-ple, the norepinephrine system has its cells of origin in the brain stem and sends ramifications throughout large areas of the neocortex (figure 9.1).

The neural arousal control systems are not only important in these early developmental stages of embryonic neural differentiation; they also con-tinue to regulate the tone of neural activity throughout the life span of neural

Figure 9.1
Major cell groups regulating cerebral arousal and their forebrain projections. In the embryo, these widespread norepinephrine influences set the activity tone of neurons that migrate from the differentiating layers of stem cells (around the ventricles at the center of the hemisphere) to take up positions in the cerebral cortex (Marin-Padilla 1998).

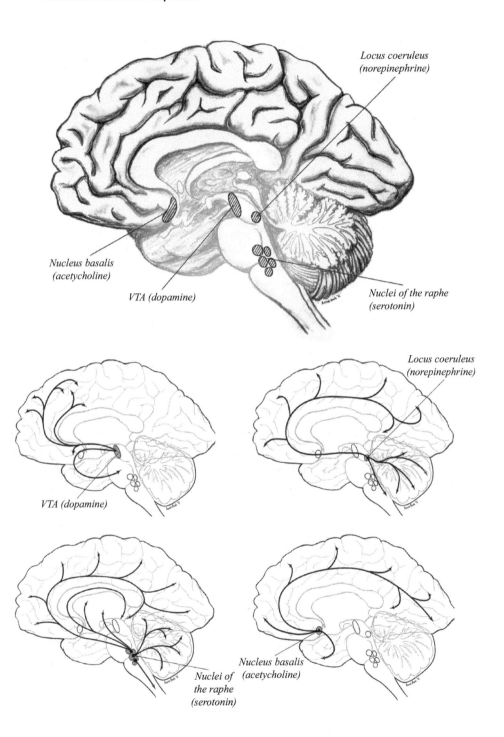

Locus coeruleus
(norepinephrine)

Nucleus basalis
(acetycholine)

VTA (dopamine)

Nuclei of the raphe
(serotonin)

VTA (dopamine)

Locus coeruleus
(norepinephrine)

Nuclei of
the raphe
(serotonin)

Nucleus basalis
(acetycholine)

development, even into old age. The developmental continuity of control of neural activity provides a useful lesson in how to reason from biological control to understand psychological control.

Norepinephrine, for example, continues to regulate the plasticity of the cortex after birth. It must be present to modulate the visual cortex during the critical periods of the young animal's early visual experience. If not, the visual cortex develops abnormally (Tucker et al. 2015). Furthermore, setting the tone of neural activity appears important not only in critical periods of childhood but throughout life. Norepinephrine modulates neural activity in ways that not only increase the arousal or activity level of its target populations of neurons, but also change this activity qualitatively, leading perceptual circuits to *habituate* rapidly to repeated inputs (Aston-Jones, Ennis, Pieribone, Nickell, & Shipley 1986). Together with serotonin, norepinephrine is integral to regulation of the habituation bias in the process of allostasis.

This habituation effect—decreasing neural activation in response to continuing input—is a unique control of neural activity in time that appears to be integral to psychological features of specific mood states. In a state of elation, for example, a person is emotionally aroused but quickly bored unless encountering new stimulation (Tucker & Williamson 1984). This effect can be explained by the habituation bias of the norepinephrine system that appears to be an integral component of the mood state of elation (Tucker & Luu 2012). Here we see multiple features of a motive control system. It regulates arousal in a way that is suited to a specific adaptive control of behavior, and it has specific influence on the ongoing cognitive process, expanding the spatial (and eventually conceptual) scope of attention at the cost of stability in time.

A different neuromodulator system, mediated by the neurotransmitter dopamine, proceeds from the brain stem (specifically the ventral midbrain; see figure 9.1) to regulate the neural development of widespread telencephalic regions, specifically those of the anterior brain (including the basal ganglia) that control bodily movements. In the neurodevelopmental continuity from embryonic differentiation running through psychological development, dopamine regulation seems to lead to increasing redundancy or *sensitization* of neural activity in time. Sensitization is the opposite of habituation in that it maintains or increases responses to constant input. This effect supports ongoing constancy in motor plans (Tucker & Williamson 1984) through a qualitatively unique control of neural activity in time. When integrated within the fully organized brain, the sensitization bias causes focus and constancy in attention as well as motor control (Tucker et al. 1995). The motivational significance of this control system is particularly important in states of threat, where fight or flight responses are carried out under the focused attention created by a state of anxiety (Derryberry & Tucker 1991).

We describe these two examples of different neuromodulator systems because they shape how we respond to certain situations and how these responses influence what and how we know. The theoretical challenge is how to reason from such neurophysiological effects to their significance for psychological development.

Perhaps at an early enough stage of neural differentiation, such as twenty-five days (figure 7.2), we would say that we are dealing only with biology, and there is nothing we would call a mind yet. But there is a point, such as in the newborn infant's early social experiences, when most of us would think that properties of mind, and personality, emerge. These emergent properties take shape within the activity-dependent organization of the brain's connectional architecture. The connections that participate in ongoing neural activity are retained; those that do not are lost. As a result, our brains change constantly, continuously tuned in their activity states by the motive controls that regulate neural activity in response to ongoing life experience.

The biology of neural development provides the basis for the psychology of cognitive development. There is no sharp demarcation between these two. Instead, there is a remarkable continuity of ontogenesis, running from the activity-dependent specification of neural connections in the embryonic brain to the differentiation of neural connections that organizes the child's mind in each new learning experience (von de Malsburg & Singer 1988). In short, the operations of mind are mediated by the mechanisms of brain development. To repeat our evolutionary-developmental mantra of natural philosophy: *cognition is neural development continued in the present moment* (Tucker & Luu 2012).

Even at the early stages of biological self-regulation, when the epigenetic mechanisms of neural control operate in an elementary, predictable, biological fashion, it may be worthwhile to recognize the implicit self—the organism—which is the reference for neural self-organization. The ontogenetic mechanisms of self-organization, and not the physical features of adult anatomy, are the active targets of evolutionary selection (Gould 1977). The selection and refinement of the developmental controls, instantiated in novel mutations of the genetic code, have constituted the key process in the evolution of each species, tuning and maintaining the physical and behavioral fitness for successful biological organisms. At least up to the present, this includes us.

The biological self is therefore the criterion for the many mechanisms of self-organization, long before the infant's neural networks evidence the first signs of consciousness. As we imagine the mind's developing process of cognition in largely unconscious form through the years of childhood, we can keep in mind the developmental coherence of the organism that has been the ontogenetic criterion for evolution. Organisms with a successful developmental program (good genes) had to maintain coherence of all their biological systems in order to grow, survive, and reproduce. This coherence of the organism—the self—continues as the criterion for successful self-regulation of each neural system.

In a general sense, the requirement for maintaining homeostatic coherence helps to explain the general architecture of the cerebral cortex that we examined in the last chapter. At the central core of the hemisphere, the limbic networks interface with the visceral nervous system, particularly with the hypothalamic monitors of the bodily state, which then regulate the motivation of behavior. As a result, the core limbic control of memory consolidation must be guided ultimately by homeostatic controls. This is one important sense in which the personal meaning of an experience relates to our survival and well-being.

At the other end of the cortical pathways, linked through heteromodal and then unimodal association areas, the primary sensory and motor networks of the somatic nervous system articulate contact with the world. These reality contacts of the somatic nervous system make it possible to evaluate our prior expectancies adaptively in light of our sensory and motor interface with our environment. Too often, psychology and cognitive neuroscience focus only on cognition and the mind's integration of the data of the world, thereby ignoring the adaptive, limbic base of cognition and concept formation. In sharp contrast, the fundamental role of the visceral function, manifested in human emotion and motivation, is essential to an accurate biological understanding of learning and memory.

9.2 From Reflex to Uncertainty

The elaboration of simple controls on neural activity in time, such as habituation and sensitization, has been a key factor in the evolution of mammalian memory and cognition. Simpler animals, including amphibians and reptiles, respond to environmental stimuli with simple reflexes or instinctual fixed-action patterns. In contrast, mammals evolved a more complex memory system that takes advantage of the representational capacity of an extensive six-layered neocortex. This neocortical mammalian memory architecture formed the biological foundation for the emergence of the human mind and its flexible representation capacity. In chapter 7 we examined the plan for the human limbic system and neocortex. That plan builds on the specialization for dorsal (spatial) and ventral (object) memory systems that is characteristic of mammals generally. These memory systems remain integral to the human mind. They provide different mechanisms for organizing cognition through the regulation of neural arousal in time and space. These mechanisms, while opaque to consciousness, shape the process of thought.

In both the rat's spatial context and the human's learning principles, the flexibility of mammalian memory allows the cumulative organization of past experience to be available to influence specific current perceptual evaluations and behavioral decisions. These are the templates of past affordances that shape present experience. However, an increasing memory capacity does not confer survival advantage if it induces a delay between a significant stimulus and an adaptive response. In the highly competitive environment of evolved ecosystems, slow responders are likely to be eaten.

The solution in mammalian evolution seems to have been the selection of those memory capacities that allow the animal to think ahead, by drawing on past experience for the purpose of anticipating impending events (Tucker, Derryberry, & Luu 2000). In the adaptive arms race of early human evolution, it was advantageous to use past experience to project expectancies about the future under conditions of uncertainty. Humans are particularly good planners.

As we will see through the rest of this chapter, the neural mechanisms for managing uncertainty have become integral to the capacities for planning and anticipation that distinguish the adaptive control of thought in the human mind. Uncertainty engaged by the need-search-satisfaction sequence is managed by the elementary motive controls of habituation and sensitization. We can use the principles developed so far in this book to understand the important continuity that proceeds from tuning neural arousal to tuning the motive control of uncertainty as the basis for knowing.

9.3   Learning Is Cognition, and Cognition Is Motivated Expectancy

The realization that mammalian learning is cognitive and anticipatory, not reflexive, is one of the most important stories of twenty-first-century psychology. Following Pavlov's demonstration of conditioned reflexes in the late nineteenth century, academic psychologists saw the conditioning process as the way to build an objective science of behavior. This approach became known as *behaviorism*. Beginning with *classical conditioning* of reflexes, achieved by associating a neutral stimulus (e.g., a tone) with an unconditioned stimulus (such as the sight of food, which causes a hungry dog to salivate), this procedure when repeated causes the animal to associate the tone with food, leading to immediate salivation when the tone is presented. In *operant conditioning*, psychologists such as B. F. Skinner took their cue from animal trainers who provided the animal with a reward when its spontaneous behavior was in the correct direction. For example, by rewarding a dog with a treat when it happened to turn to the left, the patient trainer could soon *shape* the dog to spin left in complete circles in order to get the treat.

In these classical and operant forms, the stimulus-response explanation of animal behavior was applied to human behavior in several generations of psychology textbooks. However, toward the end of the twentieth century, experts studying animal learning had observed results that were inconsistent with the traditional theory of associative stimulus-response learning. Even modest mammals such as laboratory rats were found to learn through what could only be described as cognitive expectancy (Tucker 2007).

The well-replicated results of animal learning experiments confirmed what behaviorists had successfully ignored for many decades—the realization that mammals form expectancies for the important events of their lives. When these expectancies result in successful predictions, there is no need for learning. The result is a fundamental rule that would be consistent with Dewey's notion of the practical basis of reflective inquiry. *Learning is necessary only under conditions of uncertainty.* When your predictions are accurate, you simply use them; when they are not, you learn. This is the natural extension of mammalian cognition, from the simple associative reflex (stimulus-response) memory system of primitive vertebrates to a form of anticipatory, predictive learning that is unique to mammals with their six-layered neocortex (Luu et al. 2003; Tucker & Luu 2006).

If a biological learning theory is to account for this predictive form of cognition, it must adopt a somewhat more complex organizing concept than homeostasis. Homeostasis is an adequate control process once the body has deviated from its adaptive range, such as getting hungry and then being attracted to food. Yet, mammals are more than homeostatic. They are able to *predict* the necessary state adjustment and not merely react passively. They search for food because they are interested in it, even if they are not immediately hungry. It is a rewarding exercise of their well-developed sensory, motor, and cognitive skills. Several theorists have emphasized the concept of *allostasis*—adjusting perception, evaluation, and behavior in anticipation of significant events in order to adapt to changes before they occur (Liu et al. 1997; Luu & Tucker 2003b; Schulkin & Sterling 2019). The term "allo" (from Greek) means "other" or "different," so *allo-stasis* is the process of maintaining stability in the face of variation (i.e., difference). An example of this is the way mammalian cognition serves to integrate memories to form effective *predictions*, all the while continuously regulated by the animal's needs and values as represented in visceral limbic cortex (Tucker et al. 2015). A key theoretical question concerns how these expectancies are tuned by the elementary motive controls.

9.4   The Two-Way Process of Memory Consolidation

In previous chapters we challenged the classical empiricist view that perception begins with atomistic sensory input (raw sense data) that is then combined and elaborated in higher-level association areas. We noted that this conception of perception as a one-way process has been used to underwrite the dogma of immaculate perception as well as the myth of pristine objectivity. We now know that this picture of perception is mistaken. The perceptual process is bidirectional. There is, indeed, input from our sensory interface with our surroundings, which gets combined in the various association areas of the brain. But there is also projection of expectancies from the limbic cortex, through the association areas, and out to our sensory and motor engagement with our world. Appreciating this two-way traffic in the perceptual process is essential for an adequate understanding of learning and knowing.

Until recently, the received idea was that neuroscientists had to trace visual input through the thalamus to the visual cortex, and then explain how the neuronal discharges are processed and connected to higher brain areas, such as the unimodal and heteromodal association cortices. The associations at these higher levels were supposed to allow more complex perceptions to be formed from the combination of simple ones. Strangely enough, it became apparent that the neural connections do not go only one way, "incoming" from sensory input toward association areas. Instead, there are as many projections in the outgoing direction, from association areas back to the sensory areas. At first, researchers had no idea about how to explain why higher brain areas such as association cortex would send information back down to the lower input areas such as the visual cortex. Vision, most scientists assumed, is the process of getting information *from* the eye and then making sense of it, so why should connections be going *back* toward the eye?

We now have a decent wiring diagram for the mind that was unknown to previous generations of psychologists and philosophers. What have been called the "feedback" connections of the column—from higher to lower areas—seem to organize a *predictive model* for what the visual input should be. Complementing this, what have been called the "feedforward" connections of the column—the incoming data from the sensory input—appear to propagate the *error* between the sensory input and the predictive model.

A similar pattern has been identified in other sensory modalities, so that the familiar concepts of "feedforward" and "feedback" should probably be reinterpreted. What we called the feedback or outgoing direction of traffic—from association areas out toward the sensory cortex—actually seems to *create* the initial perceptual interpretation, like a starting prediction based on prior experience. Moreover, what we assumed was the feedforward pathway is acting more like *feedback* to the initial prediction, minimizing the error between the predictive model generated within the column and the sensory data coming into the brain.

There is a well-known mathematical model of statistical decision making that fits the description of these newly discovered bidirectional neural circuit models (Friston 2018b). This is Bayesian statistics, in which the probability of new evidence is interpreted in relation to the probability set by a prior hypothesis. In fact, the meaning of evidence is defined in relation to how it fits the prior hypothesis.

Even though nineteenth-century pragmatist philosopher Charles Sanders Peirce developed similar notions of conditional probability in his logic of relations in the world (Peirce 1931), this remains an unintuitive way of thinking for the statistically uninitiated. In Bayesian logic, it is as if you can't learn something unless you already expect it. Nonetheless, it is a powerful way of handling information once you have organized expectancies for what should happen. Many neuroscientists and philosophers are rethinking human cortical processing as predictive coding in Bayesian terms (Clark 2015; Friston 2018b; Hohwy 2014).

To summarize: *There is growing evidence that human mammalian perception starts with expectancies acquired through prior experience, and then these expectancies are tested against ongoing sensory input from our perceptual interface with our environment.* We do not passively receive raw sense data that is then synthesized into unified images and interpreted at higher levels to determine its meaning. Instead, perception arises through the animal's active probing of its environment, guided by prior expectations and then modulated by error correction.

We see mostly what we expect to see. When perception does not match predictions, our neural mechanisms reconstruct expectancies in light of the discrepant evidence. This is the learning process of negotiating expectancies with reality, as seen in the behavioral analysis of animal learning experiments. Rats and rabbits are natural Bayesian learners. The cognitive process of learning is a two-way mutually interactive process, with expectations projected toward our perceptual interface with the world, and discrepancies in what is actually experienced projected back for error correction and the formation of new predictions about experience.

9.5 Knowing Favors the Prepared Mind

The predictive, expectant character of mammalian cognition requires that we abandon some of our deeply rooted views about the process of mind being built on objective perception. The new prediction-plus-testing model suggests a quite different account of perception as motivated and value directed.

What has to be given up? First of all, we can forget the dogma of immaculate perception. What we thought was the trustworthy *real data* from the eye is an error minimization process that is fully dependent on the prediction. The visual signals from the eye only become information in relation to the validity of the neural prediction that reaches out to meet them.

There goes pristine objectivity. What you expect to see profoundly filters what you are likely to see. Of course, we do not see *only* what we expect to see, since discrepant visual information can obviously rise up to challenge our projected expectancies. But research does show how strongly our prior expectations shape what we are able to see. There is a well-known experiment in which observers of a video clip are asked to count the number of basketball passes made only by people in white shirts (in a group of people with different colored shirts). Focused on this task, they typically do not see the person in a gorilla suit walking right through the group of moving passers! Once we realize that perception is not just a bottom-up, foundational process by which unfiltered sense impressions are collected and used to found knowledge claims, we have to abandon the quest for certainty that has figured so prominently in Western philosophy for the last two millennia.

In the mid-nineteenth century, the physicist and physician Hermann von Helmholtz studied both the physics of light and the brain's neurophysiology in his attempt to understand the physical basis of vision. From his reading of Kant, Helmholtz understood the mind's participation in the construction of reality through personal experience. From his studies of the physics of optics, he explored how the neural responses to light on the retina are organized to create the experience of vision. Helmholtz's intuitive integration of philosophy and physics led to the insight that the act of visual perception must involve an inference that the perceiver makes about the objects in the world (Hohwy 2014). The image that appears on the retina is often ambiguous, and requires knowledge of context to determine that it is a real object. Using this knowledge of context is an active process of inference, not a simple reporting of what is seen. Helmholtz's prescient idea seems to have resonated in the minds of several maverick—but now, through hindsight, pioneering—thinkers.

In his important article "The Reflex Arc Concept in Psychology" (1896), Dewey challenged the reigning stimulus-response theory of perception by reminding us that perception always occurs within the activity of an animal acting within its environment. Using the example of seeing a candle flame, reaching for it, being burned, and withdrawing one's hand, Dewey says that the action is primary and controls the quality of the perception:

> Upon analysis, we find that we begin not with a sensory stimulus, but with a sensor-motor coordination, the optical-ocular, and that in a certain sense it is the movement which is primary, and the sensation which is secondary, the

movement of body, head and eye muscles determining the quality of what is experienced. In other words, the real beginning is with the act of seeing; it is looking, and not a sensation of light. The sensory quale gives the value of the act, just as the movement furnishes its mechanism and control, but both sensation and movement lie inside, not outside the act. (Dewey 1896, 358)

Dewey realized that the expectant activity of the actor/perceiver to a large extent determines what will be perceived, what its quality will be, and what it will mean to the actor/perceiver.

In the mid-twentieth century, the psychologist James Gibson also emphasized that perception begins with an animal's need to act effectively in the world (Gibson 1950). The animal perceives not objects and scenes as they exist in the mind-independent world, but *affordances*, the things it can make use of. Gibson regarded affordances as adaptive opportunities of the environment that are directly perceived. Yet, he emphasized that affordances are organismic as well, the result of the sensory capacities, motor abilities, and basic needs and values of the animal as it engages its specific ecological niche. The theory of predictive coding turns out to be surprisingly intuitive for those who understand Gibson's explanation of *ecological perception*. In modern theory, at each level of the perceptual hierarchy, the neocortex anticipates affordances, as prior experiences integrate motivated actions to form adaptive hypotheses based on what the animal wants and expects in the world.

The moral of this story of the prepared mind is that the mind's process emerges first from the allostatic base provided by limbic motivational controls. In terms of neural architecture, the highest area of cortex is not the heteromodal association cortex, but rather the limbic cortex. The primary source of valued predictions is formed at the limbic level, to become tested against the qualities of the world as reflected in the prediction errors organized in unique forms at each level of the corticolimbic pathway.

In everyday thought, surprisingly enough, it may be the limbic areas of the brain—those that generate holistic, value-based, visceral feelings of situations—that are running the show when it comes to perception and knowing. Consequently, core values, motivational processes, and feelings must become the key drivers of mental operations, instead of supposedly uninterpreted sense impressions or "pure" concepts and reason. Our developing experience generates value-based expectancies that we carry forward into our encounters with our world. In animals, including humans, this process occurs at an unconscious level and operates automatically. However, humans are able to construct and become aware of plans that represent our predictions for future experience. With this capacity we can then use conscious inquiry and deliberate reflection to revise our models and predictions to accord better with present experience.

In the process of knowing, the discrepancy between expectation and what we are likely to encounter is experienced as *uncertainty* and *indeterminacy*. We naturally monitor and regulate the process of knowing through the consciousness of uncertainty. Like Dewey said, this uncertainty makes us anxious and doubtful, but it is the spur to inquiry—the search for evidence and experience-transforming action that allows the new knowledge we need to be better adapted to the world.

9.6   The Limbic Control of Uncertainty

The nature of predictive coding can be reconsidered in terms of biology as well as information theory. What are the predictions arising at the top of the top-down neocortical hierarchy? As shown clearly by the anatomy, the top of the hierarchy is the limbic cortex (figure 7.7). The visceral limbic predictions shape the error correction within multimodal sensory-motor representations in the heteromodal association cortex, which then shape the unimodal association cortex, and finally the error correction in the primary sensory cortex. The mind's architecture must be refreshed continually through the bidirectional—expecting and correcting—nature of this information traffic. The neural activity of learning and memory strongly depends on the nature of the conceptual representations (the meanings) rooted in the visceral limbic core.

But what are these visceral predictions? As they organize the perceptual process of the posterior brain, we might call them visceral *needs*. Or, perhaps as they organize action in the anterior brain, we might call them *motives*. They are visceral needs insofar as they concern the allostasis necessary for well-being, and they are motive controls of actions relative to those needs. These are primitive concepts, general predictions of qualities in the world, based in elementary homeostatic controls. To remain adaptive, the mammalian corticolimbic process must be continually responsive to bodily homeostasis, transforming the knowledge of immediate needs to support an expectant allostasis in time. As a result, the limbic states are not only reactive, but also proactive in creating predictions that support meaningful self-regulation necessary for allostasis (Schulkin & Sterling 2019). To understand the embodied mind, it becomes essential to understand how these predictions/needs/motives are regulated within neural networks, and how they are integrated within the more complex concepts of reality organized in the association and somatic regions of the neocortex.

For example, when a child goes too long without food, its uncertain expectancy is fully constrained by the need state of hunger. Normally attractive activities, such as play or socializing, no longer qualify as affordances. They are simply not perceived. The predictions that constrain the sensory processes of the corticolimbic networks are those related in some way to the experienced need for food. The actions that are primed by incipient motor predictions—motives—are similarly related to acquiring food. Anyone who has held a hungry infant knows that the meaningful expectancy is for a breast to suckle.

For needs/motives, the neural representations are no longer vague psychological constructs. They must be understood as predictive representations, concepts operating in the limbic-to-neocortical (outgoing) direction. These predictions then engage the error-correction data (incoming) at each of the four cortical levels that we have described. Only need-relevant data gets used in error assessment. At the limbic and heteromodal levels, the predictions can be considered to be integrative representations to the extent that they encompass many specific sensations or actions.

The limbic control of memory consolidation is directed by needs, motives, and general values of the self for the purpose of organizing adaptive concepts—predictions of affordances that guide experience and behavior. In his early work studying neural networks, Sigmund Freud described this fundamental mental process as the operation of the *motive memory* (Freud 1895; Pribram & Gill 1976). As we study the implications of the mammalian neocortical architecture for human experience, we can understand Freud's concept in new ways (Tucker et al. 1995).

As recognized in cognitive psychology for many years, the ways we organize memory are also ways of paying attention. Attention can be thought of as expectant memory. The scientific division between attention and memory becomes less useful as we examine the underlying neural mechanisms (Posner & Dehaene 1994). The current cognitive neuroscience of attention recognizes both dorsal and ventral corticolimbic divisions as separate attention systems (Corbetta, Patel, & Shulman 2008), paralleling the dorsal spatial and ventral object memory systems that we examined in the previous chapter.

The implication for the process of knowing is that it develops from an adaptive, motivated base. This differs from the pristine objectivity often assumed by analytic philosophers, first-generation cognitive scientists, and cognitive neuroscientists. Instead, as natural, biologically oriented philosophers, we discover we cannot understand cognition without appreciating the organismic context and its motive base.

This organismic context is fundamentally the organization of the self (Friston 2018a). Concepts are rooted at the adaptive limbic core within long-standing developmental memory structures that organize the enduring process of the self (Tucker 2007). In the process of knowing, the uncertainties of the world are conditioned by the expectancies generated by past experience and the values of the self. In strong motive states, such as lust or loneliness, the expectancies and their inherent uncertainties are narrow. The information flux of the world, and the process of knowing, are tightly filtered by the necessary affordances.

In more intellectual activities, such as reading for pleasure or listening to music, the mind's predictions may be less constrained by primitive limbic controls, but they are nonetheless motivated. What qualifies as interesting must be valued, and it is typically uncertain enough to require attention. Accomplished writers of narrative fiction understand this intuitively. They know how to play with your narrative expectations, and they know that you have to care about what happens to the characters in the story if you are going to keep reading. This determination of interest in a specific domain of information is shaped by the informed memory of one's personal history, and this interest then influences the predictive process as abstract yet distinctly valued sets of related motive, conceptual, and sensory-motor expectations, linked across the limbic-neocortical network architecture. Cognition is inherently motivated; it is an ongoing allostatic expression of the self. It is therefore not surprising just how much we humans try to make the world in our own image, both at the basic level of our organismic encounters with our surroundings and all the way up to our anthropomorphic conceptions of nature and the cosmos.

9.7 Dual Corticolimbic Pathways for Tuning Expectancy

The regulation of needs and motives of personal expectancy takes different forms in the dual anatomical pathways of the limbic system, one dorsal and one ventral (see figure 9.2). Particularly for understanding the mind's

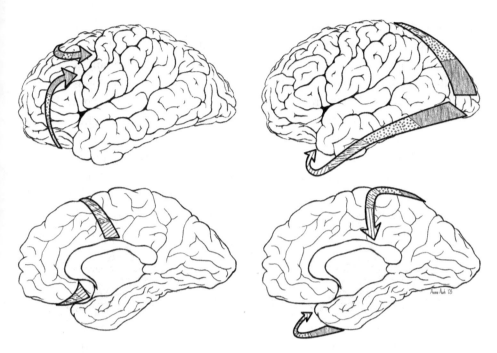

Figure 9.2
Dorsal and ventral routes for somatic sensory and motor functions of the cerebral hemispheres, seen from the lateral view of the left hemisphere (*top*) and medial view of the right hemisphere (*bottom*). For the brains at the right: Information flows into the limbic system, from the sensory systems at the back (posterior) of the brain (here shown for visual input to the occipital lobe). For the brains at the left: Information flows out of the limbic system, toward the motor systems at the front (anterior) of the brain. The dorsal route (specialized for spatial memory) is at the top half of the hemisphere, and the ventral route (specialized for object memory) is at the bottom. Arrows include markings for the different levels of primary sensory/motor cortices (*shaded*); secondary sensory/motor cortices (*stippled*); heteromodal association cortex (*dashed*); and limbic cortex, along the border (limbus) of the medial wall of the hemisphere (*striped*), using the same scheme as in figure 7.7.

adaptive base, it is important to examine the unique properties of these dorsal and ventral circuits, and their associated divisions of neocortex in the cerebral hemispheres. These specialized neural architectures support different cognitive skills, as recognized in their differing roles in spatial (dorsal) versus object (ventral) memory (Ungerleider & Mishkin 1982; Yonelinas 2006).

Theoretical analysis of the subcortical motivational circuits of these dorsal and ventral divisions of the limbic system suggests that they are regulated by specific motive controls. According to the theoretical interpretation worked out by Tucker and Luu (2012), the dorsal division appears to be regulated by the depression-elation dimension of mood control. The salience of the holistic conceptualization of the spatial context is enhanced in a good mood (elation) and suppressed when the mood deflates (depression).

In contrast, the ventral division and its focused attention seem to be enhanced by mood controls associated with anxiety and the fight-flight response (Tucker & Derryberry 1992; Tucker & Luu 2012; Tucker et al. 2000). Although still controversial and not widely accepted, this line of reasoning suggests that differential engagement of the dorsal and ventral divisions has evolved in order to shift the balance between risk and caution in handling predictions under uncertainty.

The initial recognition of the functional division of the dorsal and ventral pathways came in research on visual memory in monkeys (Ungerleider & Mishkin 1982). The dorsal pathway appeared to support memory for spatial locations, whereas the ventral pathway supports memory for specific objects. These observations were confirmed in a number of studies on mammals, including rats and humans. They were extended to reveal parallel pathways into the frontal motor cortex. Later, human studies using positron emission tomography and functional magnetic resonance imaging confirmed that both anterior motor and posterior perceptual systems are divided into the dual dorsal and ventral divisions. Additional research soon made it clear that the differential specialization for spatial and object properties applies not only for memory, but for general aspects of attention and cognition (Astafiev et al. 2003; Corbetta et al. 2008).

Consistent with the connectional anatomy (compare figure 7.7 with figure 9.2), each of the dorsal and ventral divisions includes primary sensory, unimodal association, heteromodal association, and limbic components. In fact, the specific limbic circuits of the dorsal and ventral limbic divisions are not only essential for consolidating memory in that division, but also, as we will see, they are unique enough that they could be described as dual limbic systems. Here, we find a fundamental differentiation of the neural control of perception, action, and the consolidation of memory over time. This differentiation provides clues to the ways we learn about the world.

Consider the distinctive contributions of these two neural pathways to the everyday process of thought. *Spatial* cognition allows us to understand the configurations of things, such as the spatial relations of cubicles in a business office. However, this system operates not just in spatial relations in a physical environment, but also in more abstract contexts as well. For an example from the business world, as we reason about the relations among departments in a company, we may engage configurational concepts, even though these have no concrete spatial referent. We might make a spatial map that captures the logical and functional relations among the departments (an organizational chart). This form of reasoning is thus based on the ORGANIZATION IS PHYSICAL STRUCTURE metaphor, according to which functional relationships are conceived metaphorically as spatial relations. The map then fits the spatial metaphor for relationships, defining who is "over" whom in the hierarchy. As you think about relationships, such as the relations of one department to another, your thoughts are organized in a holistic configuration by the spatial metaphor for the relational context.

In contrast, *object* cognition allows us to discriminate and separate specific coherent objects from the array of informational elements, even when these too are abstract. The definition of a perceptual object is something that retains its known identity regardless of the perspective from which you perceive it. The object pathway allows us to recognize a discrete entity and explore its properties and relations at a particular location in a larger spatial context. Objects may also be abstract, in which case we understand an abstract entity (e.g., a corporation, institution, etc.) metaphorically as a physical object. So, in our example, the inventory accounting procedure of a company may be selected as a coherent semantic object, with its own name ("first in, first out"), an enduring identity over time, and with clear boundaries that separate it as an abstract object from other business processes and operating procedures. As an object, it has an agreed upon meaning (within a social or cultural community) that maintains a tight semantic coherence in common usage.

Although the spatial and object qualities of cognition can be identified and separated, more often they flow together smoothly within the stream of ongoing cognition. The connectional anatomy, in fact, points to where they are kept separate and where they converge. The dorsal and ventral pathways are mostly separated from each other and easily identified by their unique cellular features (cytoarchitectonics). There is a greater concentration of pyramidal neurons in the dorsal division and a more distinct granular layer (layer IV) in the ventral division (Galaburda & Pandya 1983). Although these pathways are mostly separated as they traverse the hemispheres, there are small regions of heteromodal cortex, both in the posterior and in anterior networks, where these pyramidal and granular cytoarchitectonic features are clearly mingled (Eidelberg & Galaburda 1984). The implication is that these mingled heteromodal regions represent integrated, higher-order networks where the parallel predictions (and errors) of the dorsal and ventral divisions could be linked, thus providing us a sense of an object situated within its context. In simple terms, the convergence of these processing streams allows us to perceive an object at a particular spatial location.

Although it is instructive to reflect on their phenomenological qualities, the dual modes of organizing information operate largely in the unconscious background of the mind's processes. Experience involves both discrete objects that we take to be invariant, regardless of our position in perceiving them, and the spatial configuration of these elements of perception: how they fit together. In everyday cognition, these dual representational components are typically woven together so seamlessly that we are unaware that they arise from differing neural networks of the telencephalon. Let us consider the psychological characteristics of these two different neural pathways as cognition emerges from its base in motive control.

9.8   The Expansive Holism of Elation

Beginning with this awareness that cognition is organized through combinations of spatial and object pathways, we can look at the clues that these two pathways have different motivational properties that emerge from different visceral functions of the dorsal (spatial) and ventral (object) limbic divisions (Neafsey, Terreberry, Hurley, Ruit, & Frysztak 1993). The ventral limbic division seems to have evolved to specialize for the *viscerosensory* functions (feelings), whereas the dorsal division has emerged to regulate *visceromotor* functions (motive actions). These different functions seem to have different control properties that regulate the cognition of the dorsal and ventral divisions of the cerebral hemispheres in different ways. Each way may be thought of as a component of homeostasis that, within mammals, has evolved into a component of allostasis, biasing cognition regarding the uncertainties of expectations concerning future events in the world.

The dorsal limbic circuit includes the hippocampus and the closely related cingulate cortex as the core of the limbic organization of memory consolidation within the dorsal neocortex (figure 9.2). How can the dorsal limbic specialization for the visceromotor component of allostasis lead to the unique motive and cognitive properties observed for the dorsal division of the hemisphere? One interpretation is that the same form of neural control that leads to immediate visceral action creates a more impulsive and expansive form of control for the dorsal corticolimbic cognitive capacity. In cybernetic terms, this would be *feedforward* control. This form of control involves guidance *without* correction from feedback. An example of a feedforward process is a ballistic missile that, once launched, has a course fully determined by the parameters of the launch.

A related clue comes from the neuromodulators that project from the brain stem to regulate the dorsal corticolimbic division preferentially (figure 9.1). Norepinephrine is particularly dense in its projections to the dorsal limbic and neocortical regions (Pearlson & Robinson 1981). This neuromodulator supports immediate neural arousal but then rapidly habituates or decreases the arousal over time (Aston-Jones & Cohen 2005; Tucker & Williamson 1984). Such a control mode, described as a *habituation bias*, causes attention to be engaged by novel events, but it declines quickly if things do not change. The effect would be an expansive motive control of attention, orienting briefly to many elements of the environment, perhaps consistent with the contextual and spatial mode of cognition and memory in the dorsal corticolimbic division (Luu & Tucker 2003a).

How does the adaptive control of cognition serve the allostatic requirements of the organism when regulating cognition with this unique habituation bias? One theory is that the noradrenergic habituation bias is associated with the mood state of elation (Tucker & Luu 2007; Tucker & Williamson 1984). An elated mood typically accompanies the experience of success or the anticipation of success. The motive bias inherent to this mood engenders a hedonic tone—feeling good. Since it arises from prior relatively successful action, this mood state also biases behavior toward approaching new opportunities, consistent with the adaptive implication of success. Things go well, you feel good about that, and your good mood opens you to new undertakings. The habituation bias may reflect the integral cybernetic quality that evolved to regulate such a positive motive drive, inherently moderating the success response and causing it to be phasic or limited in time (Tucker & Williamson 1984). Otherwise a good mood might get out of control.

In a consistent fashion, the loss of elation, as in a depressed mood, serves to deflate the hedonic mood state, to disengage the impetus to action, and to link the negative mood state with the loss of agency within the memory of the particular context (Tucker & Luu 2007).

This influence of variation in elation on experience and behavior appears to be a basic motive control. Understanding it suggests new ways of thinking about the adaptive control of cognition. The mood state is not a *concept* in the traditional sense because its scope of meaning is so broad—reflecting the personal value of optimistic expectancy. But it may be a foundational form of cognitive control nonetheless, expanding the scope of awareness to engage the holistic context.

9.9   The Focus of Anxiety

A different form of motive control comes from considering the adaptive regulation of the ventral limbic division. As we have seen, the ventral division of the cerebral hemisphere provides the unique capacity of *object* cognition, representing the coherent perceptual elements of the environment that have enduring identities and values for the organism (Aggleton & Mishkin 1986; Amaral, Price, Pitkänen, & Carmichael 1992). The theoretical challenge is to understand how the ventral limbic division and its unique adaptive controls, including both the amygdala and basal ganglia, achieve this unique domain of knowledge.

The specific issue is how the control properties of this circuitry explain the unique physiology of the ventral division that favors cognition based on items or objects. The insula and ventral limbic network provides for control of the viscerosensory function, the evaluation of visceral input (Neafsey 1990) (figure 7.5). Perhaps in a manner complementary to the visceromotor control described above, the control properties of the viscerosensory networks provide the cognition of the ventral cortex with a unique form of regulation, in which the visceral significance of objects becomes the primary focus. In other words, these are gut-level feelings (Tucker 2002). This would be a *feedback* form of control in which the visceral criteria guide the formation of object representations in line with the visceral sense of value. The sensitization (i.e., heightened arousal) of neural activity through the ventral limbic control may then be integral to focused attention, and thus the ability to separate specific objects from the surrounding context.

The adaptive mood state most closely associated with the motive bias of sensitization appears to be *anxiety* (Tucker & Derryberry 1992; Tucker & Luu 2012; Tucker et al. 2000). This interpretation would be consistent with the strong engagement of the amygdala and ventral limbic division in response to threat, such as in the fight-flight response (Amaral et al. 1992). It would also be consistent with the focused attention that is integral to object memory in many contexts of strong motivation.

Even this cursory summary of the bodily control mechanisms underlying the adaptive basis of the limbic circuits quickly gets into somewhat dense concepts of control theory and neural circuits. Furthermore, the psychological interpretations require theoretical speculations that must be examined carefully for an effective scientific analysis. However, the general form of the emerging theory provides new ways of thinking about the biological basis for the mind's processes. Unique forms of memory consolidation—determining what is extracted from experience to become integral to the self—are associated with these major divisions of the cerebrum. Furthermore, these two modes of cognitive control seem to be motivated in unique and highly specific ways that shape the ongoing, adaptive process of thought.

In short, cognition is grounded in two motive control processes: one that provides a more holistic sense of a situation accompanied by a positive feeling of elation, and the other that provides a more focused and detailed perception of an object (within a spatial context) often accompanied by a more anxious mood. Together these two processes blend to give a sense of the value of an object in a situation relative to our overall well-being.

9.10   Tonic and Phasic Modes for Managing Uncertainty

In the last two sections, we emphasized that primitive controls on neural activity in time—the habituation of elation and the sensitization of anxiety—seem to have evolved to structure different forms of attention and memory control. These different control modes support different forms of memory consolidation and cognitive process, each with an inherent adaptive bias and each with its corresponding tuning of the uncertainty of expectant prediction. We now propose to show how these two different motivational processes lead to two different, though interconnected, modes of knowing.

Engaging the elementary cybernetics of *phasic* (transient) arousal, the mood state of elation habituates neural activity in time, leading to a broad and expansive mode of attention that quickly fades. It is phasic because it habituates rapidly, and therefore it is as ephemeral in time as it is expansive in representational scope. In psychological terms, this mode of control of the dorsal limbic circuits can be described as extraversion (Tucker & Williamson 1984), when there is little uncertainty in the process of knowing. We might call this feeling of elation the "world-is-my-oyster" mood, in which the person presses forward in full confidence and without any sense of hesitation, as long as expectancies continue to be validated by their experience.

In contrast, engaging the cybernetics of tonic activation, anxiety sensitizes neural activity in time, leading to the focused *tonic* (constant) attention of avoidance and introversion (Tucker & Williamson 1984). Anxiety both stems from and augments uncertainty. This anxious "oh-no-something's-wrong" mood is our response to the indeterminacy in a given situation. It tends to motivate us to stop, look, listen, and reconsider how we might move forward to resolve the indeterminacy, as in Dewey's description of the process of inquiry.

The neuromodulator systems regulating these mood states continue their roles in regulating the tone of neural activity and development, which begins in the process of embryonic neural differentiation. In the extension of arousal controls to regulate the continuing neurodevelopmental process, the subcortical control systems shape the brain's synaptic architecture in the lifelong regulation of neural plasticity (Tucker & Luu 2012). With this developmental perspective in mind, examining the evidence on cognitive expectancies derived from neurophysiological studies of animal learning offers new insight into the cybernetics of motive control that regulates everyday learning.

At least for the simple mammals that have been studied the most—rats and rabbits—each cognitive representation (concept) seems to incorporate both the visceral value and the features of the environment relevant to that value (Luu, Tucker, & Derryberry 1998). This also seems to be the elementary combinatory form of the concept in human cognition: personal meaning and value must be combined with the fit of the concept to the world. At the biological level, concepts are objective insofar as they arise from our sensory and motor engagement with our surroundings. Yet, they are also inherently subjective insofar as they are shaped by our deepest visceral needs.

Furthermore, in the dorsal and ventral limbic divisions, we find *two* ways of learning, each with its own motive bias (Tucker & Luu 2007). In the next two sections we consider how the dual modes of mammalian learning, implicit and unconscious as they are, shed light on the dynamic motive control of the expectant process of knowing. The central theoretical issue is how allostatic control of the cognitive process requires shifting attention in time, consolidating memory of the past, predicting the unfolding future, and tuning the subjective value of that prediction. The habituation bias of the dorsal division leads us to ignore threats and impulsively forge ahead. The sensitization bias of the ventral division makes threats loom large, so that we give them sustained, focused attention. The motive control of knowing is achieved through modulation of our elementary mood states.

9.11   Elation and Context Learning

The mood state of elation appears to be integral to the phasic arousal system and its regulation of the dorsal limbic mode of learning and cognition (Luu & Tucker 2003a). This integral alignment of phasic arousal with the dorsal limbic division could explain why positive hedonic context expectancies are so important in regulating learning, in humans as well as other animals (Luu & Tucker 2003a).

The result is a unique system of memory consolidation, gradually shaping an optimistic and holistic model of self in context. In fact, there is no strong separation of self from context. We simply feel "at home" in our surroundings and things seem what we had expected them to be. We "feel the wind at our backs," so to speak, urging us to continued activity. In this track, knowing operates under a unique form of motive control—a positive, hedonic mode of approach and forward movement. The control of neural activity in time extends not only into the past, providing a hedonic tone to memories of success, but into the future, biasing expectancy away from uncertainty and toward confirmation of valued predictions that we have learned to rely upon (Luu & Tucker 2003b).

In human experience more broadly, we could describe the holistic dorsal corticolimbic organization of cognition as a worldview. By considering the neurophysiological mechanisms we may discover the primitive neurophysiological basis for this experiential mode: extraversion. It is a way of organizing memory, implicitly and unconsciously, thereby establishing the expectancies that shape consciousness. When we expect good things, the consciousness of the world and the consciousness of self both expand and blend into a more unified holistic experience, consistent with the regulation of the dorsal division by the habituation bias of positive affect. In such cases, the world is, indeed, our oyster.

When the process of mind is consistently regulated by the habituation bias of elation, the effect may be not only a characteristically holistic scope of attention, but also a positive expectancy for success in behavioral contact with the world (Tucker & Luu 2007). In personality theory, these holistic, elation-based processes manifest themselves as extraversion. This mode of experience engages a positive affect bias that leads the person into an approach mode, engaging not only social interaction but contact with the world generally. In this state we may not learn so much, but we already know what we need to know and our elation energizes unfettered forward movement.

The specific cybernetics of neural activity in time within the dorsal corticolimbic division, habituating rapidly in the absence of novel events, is well suited to extraversion as the person orients to the novel events in the environmental context (and simultaneously habituates to sameness). Without the excitement of change, the extravert is quickly bored. Many people with hypomania (elated extraverts) become thrill seekers: the mind continually projects the hedonic expectation onto the familiar world, seeking ever more experiences that confirm their self-projections and expectancies.

As the expansive mode of cognition allows us to incorporate a wide range of information about the environmental context within the mind, the mind becomes strongly identified with the context. The valued worldview is then integral to the self. For the fully extraverted mind, the self is not differentiated from the context, but rather is isomorphic with the context, the valued worldview. This is the mode of the impulse. The exaggerated extraversion in the impulsive, histrionic, and hypomanic personality disorders reflects the imbalance in self-regulation when this mode is dominant. Both self and world become interwoven in the successful process of knowing, and uncertainty is damped accordingly. Because of the fusion of self and world, there is little or no critical perspective on one's situation.

9.12   Anxiety and Object Learning

A different form of motive control, generating a different mode of know-ing, may form the elementary basis of introversion. The ventral limbic contribution to learning has been observed in animal experiments when events, such as cues and rewards, are *discrepant* with the animal's expecta-tions (in other words, when the hedonic context model failed to predict accurately). The effect of engaging this circuit is frozen activity and a gen-eralized anxiety.

Now, if we were to empathize with the rats and rabbits in this research, we might infer that discrepant events—things you don't expect—are expe-rienced as aversive. It is important to note, however, that this is not a sim-ple reflexive response to aversive or painful events. Rather, this state reflects a cognitive process in which the animal has a clear expectancy and the events of reality are discrepant. The response of the brain's corticolimbic networks is then a cybernetic mechanism of learning, an aversively moti-vated feedback mode of adaptive self-regulation, integrating discrepant information in order to disrupt and change the context model. This second elementary process of mammalian learning begins with the first, the dorsal limbic representation of the context, and then implements a set of integral attentional and memory processes for adapting to the predictive failure of that cognitive representation. Simply put, when experience doesn't con-firm our projected expectations, we have to recalibrate those expectations and error correct, or else suffer the dysfunctional consequences.

At the primitive level, this ventral limbic mode is a fight-flight response. Yet, in mammalian cognition, these ventral limbic mechanisms give rise to a process of cognitive feedback as well, disrupting the context model (i.e., our projected expectancies) to focus attention on emergency error correc-tion to allow new learning, and thus a more accurate prediction of the new order of reality. This is learning in the sense of revising the mind's expec-tancies and not just validating them. Therefore, it has both a strong critical dimension and a strong reconstructive dimension that are absent in the elation mode of knowing. The adaptive control is integral to the learning process.

You can see why the response to a prediction failure requires the specific cybernetics of the sensitization bias. Your response cannot be phasic (like the gradual updating of the context model), but rather it must be tonic or sustained in time. As a reasonably intelligent mammal, when your world-view is suddenly proven wrong, the first evolutionary implication is that you are about to be eaten. The second implication is that you will soon starve. When the hedonic context model—your valued prediction of the world—fails, your cognition must become sustained and focused, and lead to a rapid change in learning. This is the second major cybernetic mode of the mammalian process of knowing, organized from the primitive neural arousal control of the sensitization bias. It engages defensive differentiation of the implicit self from the known context insofar as you are taken aback by the failure of your expectations and you withdraw from forward action in order to take stock of what is wrong. This generates the tonic activation of anxiety and ventral limbic control. Ideally, it also initiates a form of inquiry leading to a reconstructing of your expectations.

The findings on neurophysiology of animal learning are theoretically important, linking specific neural systems to the realization that mammals are cognitive, expectant, learners. There has been considerable progress in research on human learning as well, with the advances in neuroimaging providing insights into both limbic and neocortical networks that are engaged in handling ongoing prediction of events, and the shift in cognitive and affective process when valued predictions fail.

In humans the experience of discrepancy between expectations and actual events leads, first, to uncertainty, indeterminacy, anxiety, and doubt, and then, second, opens the way for analysis, a critical posture, and constructive inquiry to resolve the problematic situation (Luu et al. 2003). Recall, that this is exactly the structure of inquiry that Dewey described as need-search-satisfaction. This process, we suggest, is the fundamental structure of the kind of learning that can be both critical of our initial expectations (when they clash with our experience) and yet creative of a new model of experience (insofar as we can allostatically reset our expectations and models). The most elementary adaptive controls on learning turn out to be integral to the most complex aspects of human executive self-control.

The computational models of predictive control have emphasized the local computation of cortical columns, with predictions in the outgoing circuits and errors detected and minimized in the incoming circuits as described above. However, the integrated function of the mammalian brain seems to have evolved with major neocortical systems, dorsal and ventral, specialized for different modes of the adaptive control of expectancy. The greater self-regulation by elation leads to strong reliance on predictions, and insensitivity to errors, in the dorsal division. In contrast, self-regulation by anxiety leads to uncertainty of personal prediction and high sensitivity to error feedback (Tucker & Luu 2012). This is when you get the "oh-no-something's-wrong" feeling when habitual expectations are thwarted and the situation is indeterminate.

9.13   Implicit Adaptive Controls and the Feeling of Knowing

Our theoretical analysis suggests that the motive controls of elation and anxiety are not merely subjective experiences, but also cybernetic modes, ways of self-regulating the process of knowing throughout the corticolimbic hierarchy. These mood states could be described as global limbic value states—ways that we mammals organize the significant representations of life, largely in the unconscious background of experience, in the implicit neural tuning of the uncertain process of knowing.

As a result, cognition is invariably motivated and linked to emotionally significant values. This insight runs directly contrary to our received notion of knowing as motivated only by truth seeking and unfettered by values. In a neural systems analysis, we cannot separate cognition from its motivational and emotional substrates. The two primary ways of tuning uncertainty typically operate beneath the level of conscious awareness. Yet, they extend to our conscious reflection and inquiry, as in science, technology, philosophy, painting, architecture, music, dance, and ritual activities.

As we examine more closely in the next chapter, the specific form of cognitive concepts is inherent in the process of managing uncertainty. It may not be accidental that the memory representations supported by the ventral limbic system are *objects*, representations of discrete items. Threats must be identified, separated from the contextual surround, and recognized from any perspective. In this specific regulatory mode, anxiety and hostility focus attention on discrete, threatening—and hedonic-context-discrepant—objects. In human experience, this may be the motive bias of self-regulation through introversion. It is again a mode of being in the world, one in which the self must be separated and preserved from an environmental context of threat, disorientation, and frustration.

Mammalian learning thus proceeds through a kind of primitive information theory. This is not the simple mechanisms of associating reward and punishment predicted by stimulus-response theory, but rather a complementary and complex set of cybernetic modes of adaptive prediction based in elementary limbic concepts. One engages a holistic hedonic context through feedforward projection (elation and approach), and the other separates from the context, typically because the internal context model must be updated with corrective feedback (anxiety and fight-flight).

As soon as we consider the cognitive functions of these motive controls in simpler mammals, we recognize that the function of cognition is not only to know the world, but also to form concepts of the self in the world, and to act appropriately in our surroundings to achieve our ends. For most simple animals, there is minimal or no consciousness of self, so that the actions of these self-regulatory modes are implicit, unconscious. These control modes operate for us humans as well. For us, they are also mostly unconscious. Yet, sometimes they exert their effects on those dim glimpses of consciousness we manage to realize on occasion. When things are good, the adaptive mode is extraversion. You gradually incorporate the facts of the world into your holistic context model. When things are bad, you must scrap that model, break your identification with the world, and engage an introverted mode of focusing on the unexpected threats (Tucker & Luu 2012). You then have to recalibrate your expectations so that they take account of the discrepant experiences, reconfiguring yourself in the world. These are adaptive modes of being aware, and yet we have little insight into them as they operate. We simply discover ourselves with moods, and we automatically experience the world differently.

At the close of the nineteenth century, one of the first American psychologists, William James, described his own conscious experience in terms that parallel what we have inferred about the adaptive controls on mammalian expectancy and the perception of the affordances of life. In his own subjective experience, James emphasized being aware especially of *furtherances*, the felt abilities to move forward with desired goals, and *hinderances*, the checks or blocks to those desired actions. "Among the matters that I think of, some range themselves on the side of the thought's interests, whilst others play an unfriendly part thereto" (James 1890/1950, 1:299). We believe that James was astutely aware of the two modes of knowing that we have articulated from a neurophysiological perspective.

Hence, it should not be surprising that adaptive controls, mediated by the visceral limbic base of the hemisphere's networks, are integral to regulating cognition and behavior. From a subjective perspective, however, it may be surprising to discover that the process of self-control that is so important to our learning and knowing has specific and definite parameters of which we are largely unaware. We move through daily activities with the hinderances—the constraints of anxiety—and the furtherances—the impulses of elation—woven seamlessly into the fabric of behavior. These motive controls operate as subtle and unnoticed guides to the uncertainty of knowing, and we are fully unaware of their cybernetic properties. James's deeply insightful phenomenological observations could help us become at least somewhat more aware of the felt sense of hindrances and furtherances that operate automatically on the margins of our consciousness. Perhaps we can feel these motivational processes as the faint awareness of our feeling of being frustrated and blocked (leading to anxiety and doubt), or our feeling of uninterrupted flow (leading to a sense of feeling at home in our surroundings). With phenomenology guided by an understanding of neural mechanisms, we might learn to be more attentive to the feelings that define our being in the world and shape what and how we know things.

What Does This Mean for Our Processes of Knowing?

As we explore more deeply the neural processes underlying our knowing activities, it can be easy to lose sight of the forest for the trees. There is no way to avoid a certain level of detail and corresponding complexity (studying the trees) because that is what is required to explain what mind is and what actually happens in our acts of knowing.

For example, we had to look pretty closely at some of the trees of neural processes when we examined how the dual limbic modes of motive control operate within the viscerosomatic architecture outlined in chapter 8. We saw that the limbic core of the hemisphere mediates the visceral functions, such as regulating bodily mechanisms and providing the motive basis for controlling behavior accordingly. From the limbic core, motive regulation engages the heteromodal association networks, which engage the secondary (unimodal) association networks, which engage the primary sensory and motor cortices that reflect the somatic nervous system's contact with the world. These linked networks operate by forming expectancies that are projected from our visceral monitoring of our bodily functioning and well-being, outward toward our bodily (somatic) engagement with our world. This outward projection of predictions about experience is modulated via error correction in the internalizing direction from the somatic interface. This negotiation is the process of cognition through which the homeostasis at the visceral core becomes generalized into allostasis, as the cognitive capacity for learning allows the organism to extract the regularities of its world and act in anticipation of these.

There are obviously lots of trees (i.e., plenty of neuroscience details) to examine if we hope to explain our human processes of knowing. But we can get an overview of the forest, too. The general perspective that emerges from this recent neuropsychology can be summarized. What does this research mean for our lives, and why should it matter?

The answer is that our detailed account of key aspects of our motivational control systems reveals two fundamental modes of human knowing. The first mode involves a mostly unconscious projection of expectancies that are consonant with our present experience. We feel at home in our surroundings. The road opens before us, so to speak. We continue to pursue our motivated activity in a fluid and unhindered manner, so long as no contrary experience raises its head to challenge our expectations. Under such compatible circumstances, our present experience, as Dewey said, "makes sense" to us. We have a holistic, affect-rich, yet fleeting sense of the appropriate context of our actions. Our sense of self is merged with our surroundings.

Some philosophers and psychologists might balk at calling this first process a mode of *knowing* because they want to reserve that exalted status for the more conscious, exploratory activity traditionally associated with knowledge acquisition. But this unreflective process is certainly a mode of knowing, in the sense that it is a relation of a self to its world that supports its gathering of information critical to its survival and well-being, so long as it lasts. This is what Michael Polanyi (1966) aptly dubbed *tacit knowing*—the mostly unconscious, unreflective set of moods, attitudes, and practices that make it possible for us to negotiate our environment and to have conscious, focal knowledge.

The second mode arises when our projected expectancies encounter discrepant experiences at our sensory interface with our surroundings. When our habitual expectancies are thwarted, we encounter the indeterminacy of our present situation, which disrupts our purposive behavior and temporarily arrests our forward activities. We experience this indeterminacy and ambiguity as uncertainty, doubt, and anxiety. However, it is precisely this blockage of forward activity that gives rise to a stepping back—a taking stock of our situation—in order to recalibrate our expectancies in light of new conditions we have encountered. In Dewey's terms, we engage the need-search-satisfaction process, in which we come to recognize the need for reconstructed expectancies, search for a means of appropriate reconstruction, and feel the release of our energies into the world along the lines of our newly revised expectancies.

Together, the initial holistic expectancy and the response to discrepancy combine to create essential control modes for the allostatic process. The beginning of this process is the sense of being in the world where experience meets our expectations. The second phase is detecting the reality of events that may not match our expectancies. This requires a partial separation of the self from its objects of attention, in order to take a critical look their meaning for our experience. The motivational control system for this type of activity moves from the vague sense of the relevant context to a more narrow, intensified, and sustained focus on a particular object. We are then able to attend to objects, their properties, and their relations selectively as they stand forth within our present situation. Dewey called this process *signification* because it allows us to use signs (as meanings) to explore both the relations of things in the world and their meaningful relation to us.

One of our most important claims is that the same motivational processes that operate unconsciously in our embodied biological functioning are equally active in shaping our conscious higher acts of conceptualization and reasoning. In other words, there is a continuity between our sensory, motor, and affective processes as we perceive and act in the world, and those same processes as they configure our higher cognitive operations involving abstraction and language. This is one important sense in which mind and knowing are profoundly embodied processes.

The integral role of these embodied motivational controls may extend to the organization of the most complex and abstract concepts. In the next chapter we take a closer look at how abstraction is possible when the self tentatively and temporarily disengages from immersion in concrete objects of experience in order to discover new relations in the meaning of things. In the next two chapters, we take up the challenge of explaining what abstraction is, how it is possible, and what role it plays in the personal experience of knowing.

10 What Is a Concept? The Influence of Motive Control on the Formation of Concepts and Personality

It is a commonplace of traditional philosophical theories of knowledge that knowing is a process of conceptualization and reasoning that yields truth claims about the way the world is. From this perspective, genuine knowledge is assumed to require clearly defined concepts that directly refer to objects, events, properties, and relations existing in the mind-independent environment. We combine concepts to formulate propositions (with a subject-predicate structure) that supposedly represent or mirror states of affairs in the world. We express these propositional truth claims in our basic assertions about the realities of our lives, such as "My driver's side front tire is flat," "My girlfriend left me and took the TV remote," and "I know that my redeemer liveth."

As much as propositions form the substance of daily communications and of much of private thought, language is not the foundation of knowing, but rather an important superstructure built on a foundation of more fundamental meanings and concepts organized through the regularities of concrete, embodied experience. We studied examples of such embodied, experiential concepts, and how they could form more abstract knowledge, in the conceptual metaphors discussed in chapter 6. We examined the neural architecture of memory and experience in chapter 7, how learning is achieved by distributed neural networks in chapter 8, and the motive control of memory and cognition within the mind's distributed architecture in chapter 9. Our goal in this chapter is to explore one of the more provocative implications of the evidence on neural mechanisms of memory: that the cybernetics of motive control are integral to organizing the structure of concepts. Our moods, engaging the core motive controls, from anxiety to elation, determine the way mind structures concepts, ranging from high degrees of differentiation to greater integration.

Concepts are patterns of neural connectivity that constitute expectancies concerning experiences associated with particular objects, events, and actions. A concept exists as a probability distribution of expected affordances provided by certain things we experience. For example, to have a concept of a chair is to have a set of expectancies about what chairs look like, what they are used for, and how it feels to sit in various kinds of chairs. Concepts are not arbitrary symbolic *entities* held in mind and then matched with mind-independent realities. Instead, to have a concept amounts to having a specific pattern of neural connections and feeling states that arise from our ongoing engagement with the regular aspects of the physical, social, and cultural environments we inhabit.

The child's affordances are first organized around bodily needs and the importance of supportive emotional contact with the parents. They soon extend to the flexible expectancies for exercising behavioral and cognitive skills through play and social interactions. Through the many years of education, both formal and informal—mediated by the powerful semiotic process of language—human intelligence becomes increasingly refined, allowing us to partake in the rich knowledge of the culture. For a biological understanding of the embodied mind, we need to explain how the motive controls for self-regulating learning—the evolved neural mechanisms of impulse and constraint—provide the basis for the abstract concepts required for fully developed human intelligence.

10.1 Developing the Structure of Abstract Concepts through Differentiation and Integration

In a biological account of the embodied mind, conceptual systems are organized through the same neurodevelopmental process that begins with the differentiation of neural tissue in the embryo and continues throughout life with the child's education in the context of the family, school, and community. Concepts reflect the developing structure of neural connections that results from learning.

The process of embryogenesis is easiest to understand in terms of organizational structure: the neurons and glial cells differentiate in specific ways, and their functions are integrated through epigenetic (developmental) processes that maintain the coherence of the brain within the biological self-regulation of the organism. This process of neural development continues as the infant's nascent concepts become differentiated and integrated in each new learning context. The brain's genetic plan shapes only the broad outlines of cortical and subcortical organization. Each learning experience must then create the concepts that are formed through epigenetic development via the activity-dependent strengthening of synaptic connections. Our concepts are experience-induced patterns of neural connection among different functional regions of our brains. The information complexity of mind emerges directly from the information complexity of the body and its environments.

In psychological terms, the child's mind becomes more complex and capable of abstraction as her concepts become both differentiated and hierarchically integrated (Werner 1957). Differentiation means seeing differences. For example, for the young child bugs are just bugs. The child's concepts are holistic. The developmental psychologist Heinz Werner described the child's elementary concepts as *syncretic*, fusing many elements—including sensory and affective impressions—in a relatively undifferentiated, primitive amalgam of childlike experience. He described the bodily basis for this syncretic mode of thinking as the *postural affective matrix*. The bug-ness of bugs includes not only their generic visual properties and possibly such personal experiences as feeling them creeping on the skin, but also the affective properties of attraction, revulsion, and the interesting combinations thereof. Of course, these personal experiences arise in an interpersonal context, so that the child's experience of bugs is also shaped by the influential communications and actions of parents and friends that frame these experiences.

The older child's concepts become increasingly differentiated. She learns about beetles, for example, and how they can include many different varieties. The differences help to define the increasingly well-ordered complexity (abstractness) of her concepts, through the conceptual integration of recognizing that, even with the variety of forms, there are general features, or clusters of loosely overlapping features, that define the beetle. Through the neurodevelopmental process, the child's growing conceptual complexity is implemented physically in the developing complexity of her neural connections. This is the emerging structure of mind.

However, seeing differences is not enough for developing abstract intelligence. Providing order and coherence to abstract concepts requires that we understand the similarities at a more general level. These similarities must not degrade the differentiations that have been made. Otherwise, the concept would regress back to the child's syncretic holism. A concept is abstract to the extent that it applies to many different objects and events (that nonetheless share some similarities). For psychologists such as Werner, this is the core of hierarchic integration. It is a kind of integration that respects the differentiations but captures meaning at a higher level, where *higher* means that the integration is more general. Concepts that generalize to new situations—but retain their specificity and don't just become syncretic mush—are more abstract. A property or relation noticed in a present experience is seen to be applicable to other situations in distant places and times. Abstraction is then a kind of unity amid variety, where we recognize difference and yet find patterns that hold what is differentiated together in a relatively unified manner. The older child comes to understand insects, and how this concept captures a large and meaningful order of living things.

The abstract concept of insects achieves real meaningful complexity only if it goes beyond the conventional semantic definition—insects are bugs with six legs—to capture the concrete experience with many actual insects and their unique (differentiated) forms and behaviors. The semantic conventions of language are a powerful scaffold for concepts, but they are no substitute for the complex richness of concrete experience and the abstractions that can be constructed directly from it.

There are many component concepts that themselves are abstract and contribute to understanding insects. The young person who becomes interested in the biology of insects learns about morphogenesis in the early embryo, how to recognize larval forms and appreciate metamorphosis into maturity. From the perspective of cortical processing of information to consolidate memory, the young person organizes the differentiations that allow accurate contact with specific experiences. At the same time, she must work toward the coherence of understanding (accurate expectancy) that integrates the increasingly abstract knowledge.

Because cognition is neural development, the simple fact that differentiations are made implies that some functional regions of the mind's neural networks operate differently than others, and activate different functional neuronal clusters. The complementary fact that hierarchic integrations are formed implies that there must be functional relations that span the widespread differentiated regions, and these must be active when integrated concepts are operational. Our recognition of the identity of mental structure as neural structure frames the question of knowing in more explicit scientific terms—terms that would capture how relations among concepts are achieved by relations among neurons.

Within the connectional organization of each cerebral hemisphere examined in chapter 7, we saw the separation and relative isolation of specific sensory and motor cortices, as well as the high degree of interconnection of heteromodal association, and especially limbic, cortices. This architecture suggests that the *differentiation* of specific perceptual and conceptual features will be achieved more toward the sensory/motor cortices, where we can assess the discrepancy between our projected expectations and the actual experiences resulting from our sensory and motor interface with our surroundings. In contrast, the *integration* of information must be achieved more toward the limbic core, where multiple sensory and motor modalities converge. We look for greater integration of the elements of learning in the heteromodal cortex, and in the high degree of integration achieved in limbic cortex, at the central core of the hemisphere. Notice that abstraction is a graded, continuous phenomenon (i.e., a matter of more or less) rather than an either-or phenomenon (i.e., concrete vs. abstract).

By considering the psychological development of the child's conceptual systems in structural terms, we can formulate the process of increasing abstraction in terms of prediction and error correction in negotiating memory consolidation across the four linked networks of the neocortex (limbic, heteromodal association, unimodal association, primary sensory, and motor cortex). The developing brain brings its adaptive needs as inherent expectancies, and these are organized at the limbic core as the motive biases that guide learning. *Learning is shaped by the fundamental needs and values of the learner and their community of inquirers.*

These needs and motives are the engines of *integrative* predictions; they help the young mind predict and engage the adaptively significant information of the world. This is the information of specific need, as well as general interest, that becomes organized within the young brain for self-regulation (Posner & Rothbart 2000). The first conceptual integrations arise through syncretic processes in which the self is organized around significant experiences—such as being held in Dad's arms or listening to Mom's playful vocalizations (Trevarthen & Aitken 2001).

Differentiation occurs primarily through the errors encountered when expectancy meets the facts of the world as conveyed through the senses.

When, for example, a parent is distracted or unavailable, the resulting frustration leads to the child's effort to revise the expectancy in a way that fits this reality. Increasingly, the child's exploration leads to errors of inadequate differentiation that are central learning experiences. Crawling in your jammies works pretty well on a rug or carpet, but not so much on a slick floor. The error feedback immediately alters the expectancy in that ecological context, through a process of increasing conceptual differentiation. Where there was one generalized concept for crawling, now there are differentiated concepts (expectancies) for the affordances of crawling on different floor surfaces and in different types of clothing.

It should by now be clear that we are not using the term "concept" in its traditional philosophical sense of an abstract quasi-entity that supplies a feature list of essential properties that define an object or event as being of a certain *kind*. Instead, concepts as expectancies for experience are complex neural networks involved in perception, bodily movements, emotions, and feelings. Although in some contexts you can get away with thinking of these expectancies as a list of necessary and sufficient conditions for being a certain kind of thing or event, that is generally not the best way to understand conceptual differentiation and integration (Lakoff 1987). Concepts are not formed up and then stored within some distinct conceptual center in the brain. There are no such conceptual entities, and there is no such center.

Concepts are functional neural networks within and across multiple brain regions that are responsible for perception, bodily movement, and feeling. "Having a concept" is having sensory, motor, and affective expectancies associated with our experiences of certain objects and events that we come to regard as of the same kind. The expectancies are realized in the same neural networks that are involved when we actually perceive or interact with a certain object or event. There is no need to posit a unique conceptual processing center, since all of the work is done in the same primary sensory, unimodal association, heteromodal association, and limbic areas responsible for the perception, movement, and feeling that we conceptualize. Our concepts are differentiated, multimodal, and integrative—generating regularities that provide meaningful affordances for thought and action (Barsalou 1999; Feldman 2006; Gallese & Lakoff 2005; Lakoff & Narayanan, in press).

The formation and modification of concepts across cortical networks is an *embodied process* that negotiates between motivated expectancies and the evidence of one's physical, interpersonal, and cultural surroundings.

The self-regulation of this process is mediated by the strategic motive controls within dorsal and ventral limbic circuits that we examined in relation to learning for action.

The biases toward expectant impulses that we saw in examining the studies of learning are instructive in the same way for the motive control of conceptual organization. The feedforward projection of expected rewards through the dorsal limbic impulse is the initial stage for generating behavior and for generating conceptual constructions—ideas. The child's motive expectancies form elementary hypotheses for what is likely to come from the world. These valued expectancies are the primordial conceptual integrations. In the early stage, prior to the differentiations that give them richer and more complex organized structure, the child's motive expectancies may be little more than urges. But even as urges they include the agency and intentionality that will become the motive basis for dorsal limbic conceptual integration in more complex challenges of cognitive development.

In a way that mirrors the cybernetics of feedback constraint on learning—with strong emphasis on sensory guidance in the form of prediction errors—the ventral limbic influence on the conceptual process generates control by the errors of prediction that require revising expectancies. Supported by the focus of the tonic activation (sensitization) of ventral limbic circuits, the ventral limbic motive control of constraint facilitates attention to errors that are signaled by the evidence of the concept's contact with the world. The ventral limbic cortex is then able to achieve conceptual differentiation, separating individual items from their embedding in the context in the same way it achieves object memory.

The same motive controls operative in our perception of, and action within, our physical environment are equally operative, and in the same way, in structuring our developing concepts. There is a continuity between our unconscious motivated learning processes and what we call "conceptualization." Concept formation and growth turn out to be grounded on the very same motive control systems that shape our mostly unconscious learning processes. In one sense this should not be surprising, since our concepts are *of* or *about* what we learn about the world. What *is* surprising is that concepts are of the very same "stuff" as the experiences they are *about*. It is in this sense that all our concepts are profoundly embodied.

The limbic core is the site of primary integrations, in the adaptive expectancies of motivated behavior. It continually negotiates with the evidence of reality that is reflected in the more differentiated representations of sensory cortices and the specific constraints of effective actions. The dorsal and ventral limbic circuits then shift the balance between these domains of conceptual self-regulation—these modes of knowing—through their unique ways of controlling neural activity in time. We see the direct line between feeling—adaptive motive control—and knowing—achieved through specific forms of conceptual structure. The integrated result is learning, achieved within the context of our values as they bear on our survival and well-being. We also see that *knowing is for doing,* typically as an active adjustment to changing conditions.

With its habituation bias, the dorsal limbic system fosters the impulse, inclining the cognitive process toward *intention*—the expectancy of success in engaging the world. With its sensitization bias, the ventral limbic system inclines toward *attention*—processing errors when expectancy meets evidence. These elementary mechanisms of the neural process have fundamental influences on conceptual structure (Tucker & Luu 2012). When fully balanced they not only frame initial intentional expectancies, but they also differentiate these with articulated evidence. They then become organized further through hierarchic integration in a way that maintains the differentiations within abstract concepts.

Cognitive structure has been a central feature of psychological theory, in which the concept of abstract representation essentially explains how differentiations can be maintained as integrations are formed. The challenge for the present chapter is to understand fully how the structure of differentiations and integrations can be organized through the cybernetics of ventral limbic constraint (feedback control) and dorsal limbic impulse (feedforward control). We address this challenge in the next two sections by exploring the nature of impulse and constraint as they operate at two levels of self-organization: (1) as they have evolved in the control of simple actions and (2) in more complex acts of personal decision making. The relevant concepts come from control theory (Heims 1991; Wiener 1961) to explain how an action process is regulated in time. The dorsal limbic motive control is the *impulse,* an elementary feedforward control for valued expectancies. The ventral limbic motive control is *constraint,* a form of feedback control that is tightly linked to prediction errors. Each is tuned by its own motive state (Carver & Scheier 1990). The impulse is controlled by the relative certainty of elation, expecting good things from personal action. In contrast, constraint comes from anxiety resulting from the uncertainty generated by errors as personal expectancies are disconfirmed. What scientists have discovered is that the need to tune uncertainty is important not only in regulating learning and cognition, but also in organizing the neural control of the simplest movements of our bodies in space.

10.2 Motive Control in Simple Actions

We are more likely to become aware of the tonic anxious experiences of uncertainty and doubt, since they disrupt our expectancies and the forward movement of experience. We are far less likely to have any consciousness of the elation of the phasic mode because this mood state applies the cybernetic bias that confirms our unconscious expectations and predictions. As a result, when things are going our way we tend not to focus on the ongoing flow of our experience; we simply glide through the day.

Examining the unique cybernetic modes of the phases of simple action regulation—which we normally experience as an undivided unity—illustrates the opponent modes for tuning neural activity and handling uncertainty in the dorsal and ventral limbic divisions (figures 9.1 and 9.2). These motive controls have been recognized as important in regulating simple actions—movements that we perform without thinking about it, such as bringing a forkful of food from your plate to your mouth, reaching for a cup of coffee, or typing on a computer keyboard. By analyzing the neural substrates of action control, based on both clinical evidence in humans and detailed neurophysiological recordings in monkeys, we can then outline the scientific principles for understanding the control of behavior more generally. These principles yield insights into disorders of self-control that are familiar to clinicians working with brain-injured patients, but that only make sense when understood in relation to the unique cybernetics in the limbic control of actions.

The movement disorders that occur with brain damage are called *apraxias*. These are most apparent with damage to frontal regions of the brain, but they also arise from posterior brain damage, such as to the superior parietal cortex. Because of the extensive front-back interconnections of the cortex, movements organized in frontal areas rely on important contributions from parietal areas that set the perceptual context for ongoing monitoring and control of action (Luu et al. 2011; Milner & Goodale 2008).

Furthermore, as neurologists and neuropsychologists examined the motor control problems with specific frontal lesions more carefully, they were able to see a pattern that helps explain some of the anomalies of classical clinical observations. As summarized in an influential paper by behavioral neurologist Gary Goldberg (1985), this pattern reflected the differential effects of damage to dorsal versus ventral frontolimbic pathways.

The traditional literature on the neuropsychology of apraxia had long been a confusing one. In one syndrome, called *akinetic mutism* (meaning not moving and not speaking), lesions of the limbic regions of the dorsal frontal lobe result in a paucity of movement and speech generally. Given that this is the frontal lobe, which is responsible for organizing actions, it would make sense that the damage impairs movement control. However, these patients with akinetic mutism are able to move and speak effectively, but only if they are strongly motivated (such as if there is a fire alarm). The disorder seems to be not one of *ability* but *motive* for action. In fact, these patients are often misdiagnosed as depressed, and the neurological syndrome of akinetic mutism in brain damage can be confused with the psychiatric syndrome of *psychomotor retardation* in those who are severely depressed but with no known brain damage.

Goldberg recognized that the motive basis of movement—what we would describe as the impulse—was integral to the dorsal corticolimbic pathway of the frontal lobe. This is a predictive, projectional form of motor control in which the urge to move launches the action under impulsive, feedforward control. Goldberg's theoretical insight into the dorsal limbic control came from his knowledge of the anatomical connectivity revealed in the monkey neuroanatomy studies by Deepak Pandya, Helen Barbas, Marcel Mesulam, and their associates in the 1970s and 1980s. This dorsal motor pathway is shown by the top arrow at the left of figure 9.2, from the dorsal limbic (cingulate) cortex through the dorsal medial frontal lobe back to the motor cortex.

This same line of neuroanatomical research has revealed a separate frontal pathway anchored in the ventral limbic division, including the amygdala, anterior temporal lobe, and orbital frontal lobe. This pathway, illustrated by the bottom frontal lobe arrow at the left of figure 9.2, proceeds from the orbital frontal cortex through the ventrolateral frontal cortex before targeting the motor cortex. Goldberg reviewed studies that showed a very different form of higher-level control of action from this ventral pathway, in which actions are controlled in relation to the environmental *effects* of the action (rather than the internal *impulses* as in the dorsal pathway). The ventral frontal control seemed to reflect *feedback* control from the environmental data, in which the sensory data guides the action, taking over from the predictive control that seems to operate in the dorsal pathway.

This characterization of the control properties of the frontal pathways emphasizes the outgoing, predictive form of control that we reviewed in the predictive coding models in chapters 8 and 9. Yet here, predictive control seems to dominate the *dorsal* division specifically. In contrast, the incoming, error-correcting form of control (although it is engaged in each cortical pathway, of course) seems to be emphasized as the primary way of working in the *ventral* frontal lobe. We conclude that the dorsal and ventral limbic divisions regulate action with their motive control influences biased in two different ways. The bias is toward the planning and control of action in the limbic-to-neocortical or outgoing direction of prediction in the dorsal frontal pathway, whereas the bias is toward the neocortical-to-limbic or incoming direction of error correction in the ventral frontal pathway.

The specific neurocomputational operations of the linked cortical columns of the cortex that we examined in relation to predictive control can be interpreted in relation to the behavioral skills that have become specialized in the dual divisions of the limbic control of actions. The dorsal division is balanced toward impulse, as internal visceromotor urges are actualized as smoothly as possible into behavioral action. When brain damage impairs this mode of action, it makes sense that something like akinetic mutism would result. It is not the capacity to act that is lost but the motive initiative.

In contrast, the ventral limbic division supports the feedback constraint of actions through a well-developed integration of sensory feedback into the columnar organization of the ventral frontal networks. With this viscerosensory control of feedback, error correction becomes the dominant mode for guiding the process of action.

To illustrate these two motive control processes that guide relatively simple actions, consider what happens as you reach for a coffee cup. Your motor system is being regulated adaptively. You may have only the most minimal awareness of the urge to have that coffee, yet the action emerges nonetheless from its motive base. We can tell from monkey studies that this initial launch of the movement is regulated by the dorsal frontolimbic circuits (Shima & Tanji 2000). It operates under feedforward control—an impulse sent in the general direction of the cup without much sensory guidance. The dorsal frontolimbic circuits are particularly important to the large postural muscles that support the body's base of the initial impulse to action. It may not be too speculative to infer that the adaptive control of this action—mostly unconscious as it is—is associated with the positive hedonic tone of the impulse. There may be a small subjective tone of elation associated with the dorsal limbic urge to action. We can infer this because if you are in a very good mood, your impulses are particularly primed. If you are depressed, your urge to action is retarded (American Psychiatric Association 2013).

As your arm swings in the right direction, your hand is delivered to the vicinity of the cup, and the ventral frontolimbic circuits become more important. Whereas the impulse for the action can be launched with confidence (given that it just needs to go in the right direction), as the error feedback on the progress toward the goal accumulates, the uncertainty accumulates and requires a different form of motive control, engaging the ventral limbic circuits. These circuits are specialized for the fine-motor control of distal muscles, such as the lips and, in this case, the fingers. As your hand approaches the cup, you need to make fine-motor adjustments of the fingers, in order to grasp this particular cup with its specific configuration, forming the right grip and exerting the right amount of pressure to hold and manipulate the cup. This tuning process operates under strong control from sensory (error correction) feedback delivered to the well-developed granular layer (layer IV) that is found in the ventral division of cortex (Shima and Tanji 2000). The adaptive control of this influence almost certainly has its own viscerosensory feeling quality, even if it remains well below the level that engages conscious reflection. We infer that this is a quality of anxiety. The exaggerated forms of anxiety lead to high uncertainty, to the point of recurring checking behaviors in people with obsessive-compulsive disorder. More subtly, the adaptive control of feedback correction of this simple action of reaching for the coffee cup engages the ventral limbic cybernetics of anxiety to focus attention, enhance the feeling of uncertainty, and engage the fine-motor control under sensory guidance necessary for an accurate and functionally appropriate grasp of the cup.

10.3   Motive Control in Personal Decisions

Within the unconscious neural mechanisms that we rely on for simple actions, the dorsal and ventral limbic circuits thus provide different, and opponent, influences on the initiation and control of action—one facilitating the impulse initiating the action and the other applying constraint for the fine shaping of motor manipulations. Clinical evidence suggests these self-regulatory biases extend to major domains of motivated behavior and not only unconscious simple actions. Akinetic mutism syndrome implies that the dorsal limbic division provides the emotional basis—as well as the projectional control of motor actions—for the behavioral impulse. When this emotional basis is impaired, the result is a loss of the sense of agency of action. This sense of agency is subjectively important because it involves a feeling that action in the world will be successful.

The neurological syndromes with damage to dorsal frontolimbic networks likely provide insight into the neural mechanisms of the excessive impulsivity in psychiatric disorders in which there is no known brain damage. In a similar way that decreased dorsal limbic influence is seen in depression, excessive dorsal limbic influence may lead to the impulsivity of clinical mania, as when a person takes rash actions by spending money profligately or making unwanted sexual advances. The frontal regulation of action emerges from the limbic adaptive basis of behavioral control, such that the unbalanced impulsivity of excessive elation is not merely a motor deficit, but rather a deficit in personal self-regulation generally.

In a similar, but opposite, fashion, the constraint of action associated with the feedback guidance in the ventral frontal networks emerges from the unique adaptive control of anxiety. With ventral frontal lesions, the patient typically shows the *pseudopsychopathic* syndrome (Blumer & Benson 1975) in which behavior is egocentric and impulsive. This disinhibition syndrome reflects the loss of normal ventral frontolimbic constraint. As with pseudodepression, this is not the same as psychopathy, where inadequate anxiety leads the person to be unable to imagine the likely negative consequences of her actions. Instead, it is a neurological syndrome due to brain damage. Yet, there is a similar net effect: namely, the loss of the ventral limbic constraint of anxiety that normally sensitizes us to the uncertainty of our present situation and its possible negative consequences. The loss of constraint means there is impaired error correction mediated by the sensory guidance in the incoming direction of neocortical-to-limbic information traffic.

Thus, the control of both simple actions and more complex social behavior appears to emerge from the fundamental limbic basis for adaptive self-regulation in daily life. The control influences are normally balanced to the extent that we typically fail to identify them as separate controls on cognition. However, we do recognize some people as extraverts (impulsive) and others as introverts (constrained).

In the normal flow of subjective experience, we spontaneously self-regulate through the subtle currents of multiple systems of motive control, summarized somewhat simplistically here by our illustrations of elation and anxiety. Cognition, as well as action, is tuned by the perceived uncertainty of our knowledge of the world, so that we expect good things, on the one hand, as the result of the elation of forward activity, while, on the other hand, the anxiety stemming from indeterminacy and uncertainty prompts us to look out for the bad things.

The subjective quality of knowing may normally be so well balanced between these two tendencies that we don't even think of it as motivated. Curiosity, for example, appears to reflect the uncertainty of anxiety (about an uncertain situation) balanced by adequate elation generated by the intrinsic interest in the knowledge that may be gained.

Yet, there are common distortions of cognition that appear to result from the exaggerated motive biases in personality disorders. These disorders of thinking are consistent with the specific operation of furtherances (elation) and hindrances (anxiety) in shaping the structure of thought. For example, in the internalizing disorders (including anxiety, obsessive-compulsive disorder, and paranoid personality disorders), the ventral limbic influence of anxiety leads to the sensitization of vigilance for possible errors (American Psychiatric Association 2013). This motive control also supports the focused, analytic, and detail-oriented cognitive structure that suggests exaggerated engagement of the ventral limbic object memory and attention networks. These appear to be signs of exaggerated object cognition in the process of thought.

In contrast, in the externalizing disorders (impulsive, histrionic, and psychopathic personalities), there is a dominance of positive affect and strong approach motives (American Psychiatric Association 2013). These characteristics suggest inadequate attention to possible errors of personal prediction. For example, you may misread someone's facial and body gestures and mistakenly assume they are interested in you simply because you are so attracted to them. These general features of cognition in the externalizing disorders are typically marked by holistic, impressionistic thinking that suggests inadequate self-control by analytic, detail-focused cognition. The expansive, holistic cognition in these personality disorders appears to reflect exaggeration of contextual, spatial features of thought in the lives of these personalities.

Exaggerated cognitive styles in personality disorders provide important clues to how imbalance of motivational control shifts the structure of cognition in systematic ways (Tucker and Luu 2012). These effects of motive control on cognition are generally consistent with the effects of lesions to dorsal and ventral frontolimbic networks, as summarized in the classic analysis by Blumer and Benson (1975). In personality disorders without brain lesions, greater anxiety and internalizing are associated with greater constraint and error sensitivity and with more analytic cognition. In contrast, personalities associated with greater elation and externalizing manifest greater impulsivity in behavior and more impressionistic and less analytic cognition.

The remarkable conclusion is that different forms of motive control engage different and specific forms of conceptual structure. By restricting attention, the constraint of anxiety focuses cognition and therefore supports the differentiation of cognition into specific objects. This more differentiated structure is a specific form of complexity, one that is ordered by the meaningful distinctions formed by the differentiation. In contrast, by expanding intentionality, the impulse of elation supports the more holistic scope of cognition that is necessary to achieve integration of abstract concepts. To be fully organized and abstract, this integration must respect the differentiations that are maintained simultaneously. Otherwise the effect is just syncretic holism, and the complexity required for abstraction is lost. When differentiation and integration are balanced—through the dialectical complementarity of anxiety and elation—then concepts may become both complex and abstract.

To sum up, then, the same two motivational systems that appear to underlie simple actions also seem to underlie our personality formation and our cognitive activity in general. As we appreciate the implicit subjective control of simple actions by reflecting on the differential limbic control of the stages of movement, we also gain insight into the more general roles of the dorsal and ventral limbic networks in self-regulation if we reflect on the shifts in cognition and concept-formation that occur with normal mood states. In a state of anxiety, when uncertainty is experienced intensely because of a significant threat, we can sometimes discern subjectively how cognition becomes more focused and better able to delineate the objects of thought through conceptual differentiation. In a state of elation following success, however, we experience not only a sense of agency and confidence in our expectancies, but also a more integrative grasp of the big picture. However, there is a corresponding absence of critical perspective. Feelings thereby become the engines of conceptual organization.

10.4   The Unconscious Process of Adaptive Self-Control

To be clear, if we are to identify the neurophysiological controls from dorsal and ventral limbic circuits with psychological terms, such as elation and anxiety, it is important to emphasize that, in simple actions at least, these forms of motive control typically operate below the threshold of awareness. Consequently, we cannot adequately introspect these controls and their associated feeling states. It is only evidence from the effect of brain lesions on movement disorders or electrophysiological recordings during movements that allows strong inferences about the functional roles of dorsal and ventral limbic involvement in the stages of action control.

Yet, in our theoretical analysis, the evidence on dorsal and ventral limbic contributions to simple actions is illustrative of the way these dual modes of control organize the mediation between the visceral and somatic domains in higher cognitive operations as well. The limbic influences reflect the visceral roots of control. In short, urges and motives are generated in the dorsal limbic division, and constraint and error correction are affected in the ventral division. Observing the differing personality deficits with dorsal and ventral frontolimbic lesions, we saw the continuity from the elementary motive control of action to the motive control of psychological self-regulation in personality and social interaction.

Yet, this continuity must be inferred from the evidence for the very reason that it is not easily observed in conscious experience. That initial urge to launch the reach for coffee may be marginally experienced as an impulse accompanied by a small surge of elation. Perhaps you also experience a small twinge of anxiety motivating the constraint of correcting your grasp of the cup, given that this is the time when errors become most significant (e.g., without looking, you've reached for the cup, but you hit it with the back of your hand, rather than gripping it with your fingers). In normal consciousness, even if considerable correction of your impulse is required, the closest feeling you might become aware of is a mild concern over executing the movement correctly. As much as the limbic motive controls appear to be important to setting the tone of subjective experience, we are normally unaware of their operations.

This unconscious quality of motive control extends to the unconscious self-regulation of the personality. The anxious person, for example, often fails to realize that his anxiety shapes how he interprets events. Instead, he simply experiences the situation as threatening and himself as easily overwhelmed. Similarly, as the person with bipolar disorder enters a manic state, she may have only minimal insight into the way the state of elation alters her experience. She simply feels powerful and expects that the world will do her bidding. She doesn't remember a few weeks ago that, while in the depressed state, she was acutely despondent and felt worthless.

We infer the continuous influences of arousal and mood from many sources of evidence. However, the psychological principles woven into the fabric of our Western culture, as well as the science we learn with a college education, provide us with limited insight into the inherent motive biases that go with the normal mood states of daily living. We sense the mood state, and it is an important quality of experience, but we normally don't realize the inherent bias it applies on the process of cognition. Instead we blame the situation, or we blame ourselves, attributions that substitute for actual insight into the adaptive process of self-regulation.

This lack of insight works in the reverse direction, too. It supports the illusion of voluntary control over our actions by ignoring the unconscious motive controls continually at work ordering our experience and sense of self. Careful introspection often shows that, at least with the reflections available to consciousness, we are often observers rather than deliberate agents of our decisions. In his effort to integrate the new science of psychology with his own subjective experience, William James reflected on the limitations of conscious insight into self-control. He described his experience with the nature of personal will in his efforts to get out of bed on a winter morning (James 1890/1950). Living in the era before central heating, James found himself forming the conscious intention to get up and leave the warmth of his covers. Despite his conscious and deliberate exercise of the will, nothing happened. Instead it was only when his consciousness was distracted, such as by remembering a meeting he should attend, that he found himself acting out the impulse to get out of bed and start the day.

10.5   Mind Emerges from the Concrete Unconscious

At a basic level, both acting and knowing are motivated, even if we are typically not conscious of *how* our motive controls operate or even *that* they shape our experience. The motive influences are primitive and elemental aspects of concrete cognition. They operate at the simplest level as we manifest the motive urges of valued expectancies and then adjust those expectancies in light of discrepancies with our actual ongoing experience. Even though it is largely opaque to introspection, we can infer the biology of cognition at this elemental level and appreciate how it becomes organized toward greater conceptual complexity.

Concrete thought is *stimulus bound*, meaning that it is fixed in relation to a specific stimulus or environmental context. It is not unlike a reflex, in that the mind is fused with a specific perception or action. Abstract thought, in contrast, captures more general features of the world at the same time as it allows differentiations between unique items or contexts.

We can observe this process of emerging abstraction by studying the developmental progression from the concrete thought of the young child toward the increasingly abstract thought of the young adult. As Swiss biologist and psychologist Jean Piaget studied children solving problems, he noticed that they understood problems holistically. For example, when shown two differently sized jars of water and asked which has more water, the young child understands only one dimension, such as the height of water in the jar, and ignores the other dimension, such as the width of the jar. Piaget described this concrete mode of experience as *centration*, forming a concept on one central dimension. Concrete thinking of this sort is holistic in the same way as Werner's notion of *syncretic* cognition. It fails to include the conceptual differentiation required to consider multiple separate objects or features of experience simultaneously. Conceptual differentiation is necessary to consider both the height and width of the jar in order to determine which jar has more water. The process of conceptual differentiation required to overcome centration is a key contributor to increasing abstraction and conceptual complexity.

For Piaget, the child at this stage, preschool or kindergarten, is not only concrete but also egocentric. Egocentrism is a kind of centration or syncretic experience in which the lack of differentiation fuses the self with the situation it experiences. Knowledge is only from an egocentric perspective.

This co-occurrence of the child's syncretic, undifferentiated thought with his egocentrism was a key insight for Piaget. It implies that the greater conceptual differentiation required for complex concepts applies in the relation of self to world. When thought is concrete, so is the relation of self to world, and each experience is unavoidably egocentric. The key insight is that concepts are not primarily about relations of objects in the world, but about relations of the self to the world. Egocentrism and holism are integral aspects of concreteness because the self is not separate from the idea. Concepts are predictions of the state of the world that are also statements of the knowledge of the self. Because concepts are relations of the self to the world, as the concepts become more differentiated, so too does the relation of self to world become more differentiated (Harvey, Hunt, & Schroder 1961).

We recognize that the challenge for conceptual differentiation is embedded within the holistic context of the child's experience. The understanding of self and world for the young mind is concrete and syncretic, such that the features of nature that are learned and expected are an important part of being at home in the world. The expectancies of the regularities of the world are integral to the organization of the self. A challenging conceptual differentiation challenges the old self and requires us to become someone new via reconfigured expectancies.

The increasing organization of the developing child's neural networks supports the increasing complexity and flexibility of experience and, consequently, of the self. The integrity of these neural networks is a requirement for maintaining abstract thought throughout life. This specific dependence on healthy neural networks is shown by the increasing abstraction in the child's cognitive development and, conversely, by the loss of abstraction and the restriction to concrete thinking that occurs with brain damage.

Mind Regresses to Concreteness with Brain Damage

Although the frontal lobes are particularly important to abstract thought, neurologists and neuropsychologists have observed that almost any form of brain damage leads to more concrete thinking. The implication seems to be that abstract concepts are not easily localizable to a specific brain region, but rather they require the integrity of widespread networks to organize the complexity through both differentiation and integration of the distributed networks. When this neural complexity is lost through brain damage, the result is the loss of abstract thought.

In the traditional neurological exam, the physician asks the patient to explain what it means to say that a rolling stone gathers no moss. The abstract meaning of this idiom—about moving and attachments—requires going beyond rocks and moss. A person with brain damage, or serious psychopathology such as schizophrenia, often loses the capacity for abstraction and will not be able to explain the idiom. Of course, the clinical assessment requires information on the person's previous educational and intellectual level in order to determine if there is an actual loss of a previously intact function. But the theoretical implication of the concrete thought in brain-injured people is clear: the more complex form of knowing required for abstract concepts requires an intact brain and thus the ongoing complexity supported by widespread neural networks.

The neurologist Kurt Goldstein (1939) studied veteran German soldiers recovering from gunshot injuries after World War I. Injuries that damaged abstract thought were typically to the frontal lobes, the uniquely human integrative zones. Goldstein would ask, "What if the snow were black?" The person able to think abstractly would be able to entertain this strange question with some possible deductions, such as how dark things would look or how snow would melt faster in the sun. But the person with a frontal lesion cannot imagine such a thing. He will insist that snow is white. The idea cannot be separated from the self and manipulated flexibly, but rather is fused with the immediate, egocentric apprehension of experience.

Goldstein's description of his work with the brain-injured soldiers is significant because he recognized the debilitating effect of losing higher cognitive skills. He got to know these men over many years and appreciated that cognition was not an isolated feature, but rather a function of the organism, the whole personality (Goldstein 1952). Goldstein introduced the term "self-actualization," which became a central idea in humanistic psychology to explain the most well-adjusted and broadly intelligent personalities (Maslow 1968). Goldstein first applied this term to describe the organismic basis for neuropsychological function of his brain-injured patients. As he observed the daily behavior of these rehabilitating soldiers, he saw how their coping efforts continued to reflect the inherent motive to maintain and develop organismic integrity, their identities.

Instructive evidence of the disordered self has come from specific lesions to the dorsal limbic base of the frontal lobe. Dorsal limbic lesions may lead to the loss of initiative and to the appearance of pseudodepression. The importance of the subjective awareness that goes along with this dorsal frontolimbic activity is shown by another neurological syndrome. Patients with extensive mediodorsal frontal lesions (such as from a stroke or tumor) show unusual reactions that may illustrate the effort to make sense of the world when the agency of the self is disrupted. An example is the *alien hand syndrome*, in which the hand opposite to the dorsomedial lesion (e.g., the left hand following a right hemisphere lesion) is experienced as acting on its own. Because it has been disconnected from the dorsal limbic impulse that forms the primitive basis of agency in executive self-monitoring, it is as if the action is not part of the self (Goldberg & Bloom 1990). This effect may not be restricted to the hand. One patient was found on the floor of his hospital room. He explained that he woke up to find a strange leg in his bed so he grabbed it and threw it (and himself) on the floor.

If we think back to the previous account of the frontolimbic control of simple actions, we saw that the dorsal limbic influence leads to the impulse to launch actions into the world. Here the evidence on the loss of the sense of ownership of actions with dorsal lesions suggests an additional perspective. The self as agent is the implicit basis of the impulse toward action. Although we have only a semiconscious awareness of self, it forms the sense of agency that allows us to recognize our actions as our own.

Even when only minimally conscious, this sense of agency is an integral motive control of experience. It can be seen in the exaggerated dorsal limbic contribution that appears with the abnormal elation in clinical mania. The neural arousal control that we call "elation" leads to an exaggerated sense of personal agency that is reflected in grandiose ideas, rash decisions, and impulsive spending of money.

This syndrome appears to be another reflection of the unconscious influence of motive controls. The sense of agency—that your behavior is a result of your intentions—is an implicit but integral control in normal consciousness. When it is insufficient, we may become paralyzed with depression. When it is excessive, we may become pathologically impulsive and grandiose. And when it goes missing altogether, we may find a strange leg in the bed.

Another striking effect of loss of the sense of personal agency is that we may conclude that someone controls our thoughts. This effect is often seen in the cognition of people with schizophrenia. They often show a loss of the abstract attitude, with cognition that is concrete and disorganized, appearing bizarre to someone who doesn't understand the person's effort to make sense of their disordered experience. A common feature of schizo-phrenic thought is the delusion of external control, in which the person thinks his mind is controlled by some external influence, such as aliens, the CIA, or voices on the radio. One way to interpret this disordered think-ing is as a kind of concrete thought that is due to a specific impairment of the dorsal frontolimbic circuits that normally contribute the sense of per-sonal agency to behavior. With this function impaired, and thus without the normal sense of being in control of their actions, the schizophrenic person observes their own thoughts and behavior and feels as if someone else controls them (Blakemore & Frith 2004).

The principle to be derived from these examples is that the personal agency of normal dorsal limbic control is an integral mechanism of consciousness, with specific influences on both subjective experience (being in control) and the structure of cognition (integrating concepts). In these pathological examples, egocentrism appears to be a concretizing influence, degrading the capacity for abstract thought in direct proportion to the degree of egocen-trism. A manic person may be highly creative in moderate degrees of the elated state but as the elation is exaggerated, his ideas become disorganized and inappropriate in direct proportion to becoming personally grandiose.

Yet, the generative process of knowing may depend on an adequately elated mood state in the best of circumstances. People who are creative writers have a propensity to develop affective disorders (mania and depres-sion) (Andreasen 2008; Andreasen & Canter 1974; Andreasen & Powers 1975; Shaw, Mann, Stokes, & Menvitz 1986). The sense of agency and the expansive thought associated with a strong dorsal limbic mode of self-regulation may yield creative work, generating flexible ideas (expectancies) for what the world might be like. Moreover, those who also experience a good deal of anxiety (about their presumed creative insights) may focus on exploring and developing a creative thought in an almost obsessive-compulsive reworking of the art object, such as multiple drafts of a poem or novel or multiple versions of a painting or musical work. Painters, for example, may work on an idea by producing scores of attempts to express more adequately and fully what irritates them and cries out for expression (e.g., Monet's many haystacks or his water lilies, Picasso's analytical cubist faces, or Georgia O'Keefe's flower series).

The requirement for creative productivity may be the ability to maintain some balance between the creative urge and the ventral limbic influence of constraint and differentiation, so that creative generation remains grounded in the evidence supplied by our experience. Otherwise, the generative energy of the dorsal limbic impulse becomes a concretizing influence, degraded by its inherent egocentrism, rather than a productive component of integrating abstract insights into the real world.

It may seem that abstract thought is maintained in the anxious, obsessive, and paranoid personality disorders. The cognitive style of these people is often described as *intellectualized* (Shapiro 1965), so it might seem as if intellectual capacity is spared. However, the rigidity and restriction to literal interpretations indicates that the integrative capacity of generative cognition is substantially diminished. These people whose personalities are crippled by excessive constraint are unable to engage the flexible and creative generation of ideas required for truly abstract thought in everyday creative problem solving.

The argument of this section can now be summarized as follows: the dependence of cognitive structure on self-regulation that we have inferred from the developmental evidence can be seen not only in the abnormal cognition of people with brain lesions but also in the psychopathology of mental disorders, including both severe psychoses and less severe personality disorders. We see the normal patterns of motive control of concepts, with greater impulse and intention supporting conceptual integration, and greater constraint and attention supporting differentiation. Adaptive motive controls directly structure the differentiation and integration of cognition. Yet, these motive control influences seem to become distorting exaggerations of mind when one mode is adopted as a coping strategy at the expense of the other.

In each case, unbalanced self-regulation seems to lead to both egocentrism and concreteness of thought. Once we recognize the cardinal forms of the limbic modes of motive self-control, with the structural influences they have on cognition, it becomes clear how excessive limbic influence has a concretizing influence on conceptual organization. Even though the greater interconnectedness of limbic cortex implies that this must be the integrative zone for cortical networks of cognition, the limbic base of concepts appears not to be abstract per se, but rather *syncretic*. Fused with the homeostatic controls from subcortical circuits of the limbic system, the limbic cortex can be seen as adaptively integrative, but only in the diffuse, holistic, and poorly differentiated sense of syncretic cognition. Abstraction seems to require integrated function of the full corticolimbic network system.

The disorders of personality can thus be understood as disorders in the regulation of the self in relation to the social surroundings. The self often becomes a concretizing influence, implying that some degree of self-actualization may be required to achieve and maintain abstract cognition. In self-actualization, the needs of the self are resolved, more or less (Maslow 1968). The process of knowing proceeds under more optimal forms of motive control that support abstract concepts rather than degrade them. Otherwise, the process of knowing is distorted by the search for knowledge that would support an unrealistic, narcissistic self, thereby degrading the capacity to understand the world. When this occurs knowing cannot be objective, but each thought must be a defense of the self. In the narcissistic personality, each event of daily experience is interpreted as either a threat to the self-image or an opportunity for self-aggrandizement, causing the person to be insensitive to the objective nature of ongoing events, thereby losing effective contact with reality.

The implication of studying exaggerated limbic controls, with their unique structural influences, is that abstraction requires a balance of both differentiation and integration, with the balanced self-regulation of attention and intention supporting these achievements of conceptual order. Even as the capacity for abstraction is essential to the more advanced forms of human intelligence, it is a fragile capacity. It depends within each moment on the physical substrate of neural networks, and within each life on the social and cultural context in which these networks of the personality develop.

Just as abstraction seems to require the balance of differentiation and integration, concreteness appears to be the default state of mind when this balance is not achieved or when it is degraded for any reason. The balance seems to require that the motive controls are subordinated to (and engaged in support of) the conceptual process, rather than being driven solely by personal need. Yet, in each of the examples of intelligence in the real world—in the optimal achievement of abstract concepts or the degraded egocentrism of concreteness—the process of mind is invariably a subjective process. The mood states, such as elation or anxiety, clearly change the quality of subjective experience. However, they also exert specific influences on the structure and process of cognition. Elation expands cognitive scope at the same time as its habituation bias causes the cognitive process to become impulsive and ephemeral. Anxiety restricts cognitive scope as it constrains neurocognitive activity to become focused and tight. Along the way, these motive controls of impulse and constraint intrinsically regulate the experience of uncertainty.

As knowing emerges from the subjective foundations of neural control, the implications of recent theoretical work seem to support Kant's (1781/1968) insight that the process of knowing is as much a kind of self-expression as it is a process of apprehending reality. With this realization we can return to the questions of chapter 3, recognizing that the process of knowing is inseparable from the motive controls that are integral to self-regulation, providing the sense of agency that allows conceptual integration, in contrast to the uncertainty and trepidation that signals differentiation and constraint. Human cognition is found to be an invariably adaptive process, not entirely conscious but controlled in each moment by its visceral foundations in emotions and motivations. The mind's elementary mood states apply their specific influence on the structure of experience. As the attention of anxiety weaves the articulated differentiations of an accurate representation of reality, the intention of elation reaches out to generate the integrations of meaning that link the self to the world.

10.6   The Semantic Scaffold

We have emphasized general properties of cognition that become more abstract with increasing structural organization. Most of this cognition goes on beneath the level of conscious awareness as we acquire meaning through our engagement with our surroundings and our social interactions. Our gradual acquisition of our language depends on all of this nonlinguistic meaning making, but language opens up new possibilities for meaning and thought not available to the prelinguistic infant. So, our experience of meaning extends below and beyond our linguistic development, and then that meaning is enriched, deepened, and expanded as we grow into language and signification. Through learning language, each child gains ready-made abstractions—often in metaphoric form—that convey the conceptual system of the culture. Practice in understanding the meaning of speech provides basic exercise in marking important distinctions in our experience, recognizing relations among objects and events, and interacting cooperatively within our social communities. For the literate child, the medium of language opens vast horizons of meaning, leading to abstract knowledge.

Nonlinguistic cognition reflects a fairly direct interaction between the visceral and somatic domains we described in neuroanatomical terms. Through our bodily transactions with our surroundings, we develop concepts that retain the immediacy of sensory and motor experience.

In contrast, the acquisition of language makes it possible for us to go beyond immediate experience, to discern relations to objects and events distant in space and time. In his pragmatist logic, Charles Sanders Peirce proposed that language provides the *semiotic* function, representing features of the world with signs that depend not only on the referent in the world, but also the interpretation that the person brings to understanding the meaning of the sign (Wiley 1994). Through the regularity and systematicity of language, humans gain a remarkable skill for understanding organized complexity through the abstract concepts of the culture.

Linguists have long distinguished between the *syntactic* aspects of language (the formal grammatical structures), the *semantic* aspects of language (the shared meanings), and the *pragmatic* aspects (the uses to which meanings are put, and the practical meaning of the language in a particular context) (Givon 2002). Consistent with the traditional search for foundational literal meaning that we considered in chapter 2, semantic definitions are often seen as the basis for objective meaning, whereas the pragmatics of the language act are seen as mere practicalities of the specific speaker in a particular setting and speech community.

In sharp contrast with this received view, the radical separation of semantics from pragmatics cannot be sustained in light of how we actually generate meanings and communicate with others. As we have seen in chapter 6, meaning emerges only through our bodily engagement with our surroundings (including social relations with other people), and this meaning is always embedded in, and shaped by, a broader context rooted in our deepest biological and interpersonal values. Moreover, the syntax and grammar of a language are not just matters of pure form. Instead, syntax is *meaningful form*—form that arises from patterns of our meaningful engagement with our environment (A. Goldberg 1995; Lakoff 1987; Langacker 1986). Both syntax and semantics arise from the bottom up (out of our bodily engagement with our world) and not from the top down (out of some alleged realm of pure form or disembodied concepts) (Johnson 2007). Just as syntax and semantics are embodied, so pragmatics—the use of language within a speech community to perform certain communicative acts—is grounded in our embodied capacities for action. Consequently, syntax, semantics, and pragmatics are all inextricably intertwined and arise from our bodily interactions with our surroundings (Feldman 2006; Lakoff 1987; Lakoff & Narayanan, in press).

Meaning is not essentially, or necessarily, a linguistic process. We saw in chapter 4 how Dewey (1925) described the vast continent of embodied meaning (which he called *sense*) that precedes and underlies our linguistic expressions of meaning through signs (which he called *signification*). Meaning that is expressible in language rests on a far more expansive process of nonlinguistic meaning making that operates mostly beneath our conscious awareness. This meaning arises from our embodied engagement with our world, under motive controls of elation and anxiety that shape and regulate learning and conceptualization. This embodied meaning operates in generating linguistic meaning, but it also underlies all of the other nonlinguistic meaning processes involved in symbolic communicative activities such as painting, sculpture, architecture, music, dance, theater, spontaneous gesture, signed languages, ritual practices, and other forms of embodied symbolic interaction (Johnson 2007, 2018).

Even as he devised a logic to articulate the information concerning the relations of things in the world, Peirce's formulation of the semiotic process made it clear that signs (words and other objects and actions) are linked to reality in an active process of interpretation. Signs can be meaningful, not only because they sometimes refer to mind-independent states of affairs, but, more importantly, because they indicate possible experiences for a person. The signifying relation is tripartite: a relation between a sign, the experiences it indicates or enacts, and the person who grasps the meaning of those experiences. The sign is meaningful only for an interpreter and only relative to context.

Recognizing that all conceptualization, reasoning, and knowing are situated, selective, and partial, Alfred North Whitehead (1938) observed the tendency for scientists (and people in general) to mark certain distinctions (objects, properties, relations) embedded in their language and then to define the elements of reality (semantics) solely in terms of these distinctions as the facts of their discipline. They proceed as if the true meaning of things has been adequately captured in their preferred vocabulary or conceptual framework. Whitehead christened this the *fallacy of misplaced concreteness*. Similarly, Dewey called this tendency the *philosopher's fallacy*, where meaning is reduced to the semantics of one's favored philosophical or other theoretical description. We realize therefore that even as it provides a powerful scaffolding for organizing and communicating knowledge, language does not capture the full range of operative meanings, nor does it alone generate abstract concepts. The challenge of developing abstract complexity through fully embodied cognition remains essential for personal participation in the process of knowing (Bergen 2012; Feldman 2006; Lakoff & Narayanan, in press).

10.7   The Continuity of Neural Control

We have suggested that for both verbal and nonverbal forms of cognition, there seem to be regular and systematic changes in conceptual structure that are associated with the strong motive biases. These biases are particularly apparent and systematic in personality disorders, but they can be seen in the basis of dorsal and ventral limbic controls of simple actions as well as more complex cognition. Such effects indicate an inherent influence of motive control on conceptual structure.

The more impulsive and extraverted modes of self-regulation seem to foster a more impressionistic and holistic integration of conceptual structure. These modes suggest the exaggerated influence of the dorsal frontolimbic networks. The more constrained and introverted modes of self-regulation seem to favor more analytic and focused forms of cognition, suggesting greater differentiation in conceptual structure. Both the cybernetics (constrained control) and representational structure (differentiation) in anxiety disorders seem to reflect greater involvement of ventral limbic influences. There appears to be a strong continuity running from the mode of control of neural activity in time to the organization of conceptual structure. These two motivational control systems operate, as we saw, at the level of sensory processing, simple actions, personality formation, and cognitive processes, so there is a motivational continuity among these four dimensions of our selfhood.

Although the exaggerated forms of motive control and conceptual structure in the personality disorders and psychopathology are instructive for understanding these intrinsic relations, the more general implication is that representation and motive control are integrally bound together in normal cognition as well. If this is true, then the ability to organize complex conceptual structure with both differentiation and integration requires the appropriate self-regulation of anxiety and elation, respectively, as part of the ongoing process of achieving and maintaining conceptual complexity. As the child learns to organize concepts to span the information of the world, she must unconsciously engage the motive controls of constraint and impulse to achieve the necessary organizational structure. These motive controls are given different names in the context of intellectual effort, such as discipline, initiative, or creativity, but they reflect the basic operation of motive controls, together with their inherent mood states, that appear integral to organizing conceptual structure.

The conclusion of this line of reasoning is surprising, and it challenges some of our most deeply rooted assumptions about mind and thought. *Organized and thus abstract and complex concepts are not the product of pristine objectivity, kept free from the taint of emotional and motivational influence. Rather, motive controls are the essential engines of tuning neural activity in time to weave together the differentiations and integrations required for the structured, abstract concepts that make it possible to understand the complexities of reality.*

In this chapter we emphasized the inherent influences of the dorsal and ventral modes of motive control as they operate more or less unconsciously. The examples of exaggerated cognitive structure—either focused, tight, and analytic or diffuse, emotional, and impressionistic—remain concrete when they appear in the thought of those with personality disorders. How fully integrated abstract forms of conceptual organization are achieved in optimal development of mind will require that we consider, in the next chapter, the role of self-awareness. We return to the questions of chapter 3 about how knowledge is organized on the foundation of the self. This will lead to an instructive observation on human development: the emergence of mature self-awareness seems essential to the simultaneous development of abstract thought and conscious self-regulation in maturing adolescents. This integration of abstract thought and self-aware control appears to be essential for developing the mature and deliberate capacity for human knowing.

11 Abstraction, Self-Awareness, and the Subjective Basis of Knowledge

The capacity for abstraction is almost nonexistent at birth, but it develops as we grow into adolescence and become more self-aware. As it develops, we become more and more capable of the abstract thinking characteristic of our highest level of cognitive achievement in the sciences and arts. It is a long journey from knowing how to walk or knowing what a dog is to knowing how to solve equations or knowing how nuclear fission works. Yet, there is a continuity among these varied cognitive activities, which are all rooted in the memory capacities that depend on the brain's motivational controls. By organizing our motive controls in the service of constructing abstract concepts, we come to know the complexity of the information in the world. In this process, we gain a corresponding complexity of experience and self-identity.

In this chapter, we will explore the implications of this naturalistic account of mind. Chief among these implications is that knowledge is an *organismic* process, a product of the whole person. The same motive controls that organize the developmental form of our personalities—whether normal extraversion or introversion, or the more abnormal histrionic, impulsive, or obsessive-compulsive personality disorders—are the means by which we construct the concepts that form our knowledge of the world.

In our modern philosophy of the organism (Whitehead 1929/2010), mind is emergent from subjective, inherently motivated experience. How, then, is objective knowledge possible? One answer is the discipline of education. We are schooled in using our conceptual capacities of analytic and holistic memory organization so that they are not only expressive of our needs, hopes, and fears, but also responsive to the error correction of contact with the world. But there is a more complex answer that comes from understanding the way that adolescents overcome the egocentrism of child-hood to achieve the abstract thought of the adult. As the philosopher Fichte first realized, we become self-aware and less egocentric through the process of intersubjectivity as we learn to appreciate the perspective of significant others in our lives. As a result, our natural philosophy must extend from its roots in the biological mechanisms of the brain to span the personal experience of culture. We learn to become objective through overcoming egocentrism through the exercise of intersubjectivity. Fully understanding someone else's perspective teaches that knowledge is not just what appears subjectively, even if that is the fundamental perspective of mind.

These are then the roots of objective knowledge: our fundamental egocentrism of motive control grows to maturity through formative experiences with intersubjectivity and through education, both formal and informal. Although the emotional and interpersonal roots of mind are eschewed by those who assume pristine objectivity, these roots are essential for understanding how objective knowledge emerges in the world from its subjective basis. After considering how everyday thought arises from the embodied basis of personal experience in both successful and unsuccessful personalities, we will examine how these same roots of thought make scientific knowledge possible. In scientific inquiry, both the subjective basis and its dialectical exchange with intellectual culture have been shown to be essential in the progress of abstract objective knowledge.

11.1   Abstraction Is the Process of Concept Formation

We have been arguing that the process of conceptualization *is* the process of abstraction. A concept captures the meaning of many specific experiences in a way that generalizes to new situations. This generalization requires abstraction, so that the concept is not tied to specific, concrete experiences. The abstractive process involves recognizing individual features that nevertheless remain integrated into a general recurring pattern of experience. As William James (1911) put it, in concept formation, we select a feature or pattern present in the "much-at-onceness" of a particular perceptual experience and recognize that this feature or relation is also manifested in other objects and events, thereby achieving a general character not tied to any particular percept. We grasp the meaning of that feature or relation by virtue of the expansive semantic role it plays in many other experiences.

Hence, meaning grows, deepens, expands, and is enriched by new, more broadly applicable relations.

There is a long-standing tradition in Western philosophy and psychology that distinguishes so called concrete from abstract concepts. Concrete concepts, on this view, are directly tied to sensory and motor functions. They are concepts of things we can see, touch, manipulate, hear, and taste. Abstract concepts, on the other hand, are regarded as independent of sensory–motor experience. Concepts such as *mind, thought, love, democracy*, and *freedom* are abstract because their meaning cannot be reduced to sensory and motor perceptions and actions.

Although this concrete-versus-abstract dichotomy is a workable distinction for some occasions and purposes, it can be quite problematic in others. There are contexts in which it is useful to distinguish between perceivable objects and realities not reducible to perceivable physical characteristics. However, when this concrete/abstract distinction is reified into a rigid and absolute epistemological and metaphysical dichotomy, it becomes difficult to see what concrete and abstract concepts have in common as concepts. Even worse, it purveys the mistaken assumption that concepts and percepts must be processed in fundamentally different regions of the brain using different neural systems.

On the view we are developing, the process of knowing begins with concrete experience and becomes abstract as it becomes more fully organized and generalizable through the motivated operations of differentiation and integration. There is no definite threshold between concrete and abstract, only increasing organization of the memory representation. As concepts become more complex and organized, they become more abstract, respecting the particularity of the diverse objects or events through differentiation while grasping their commonality through integration. That is why conceptualization *is* abstraction. Concrete cognition is holistic and undifferentiated, a level of knowing that is fused with immediate feelings, sensations, and urges. In this form, experience is fundamentally egocentric. It is then abstraction that allows us to move beyond the fusion with, and immersion in, an immediate experience to see relations with other experiences, objects, and contexts, thereby deepening and enriching meaning, bringing objectivity to knowledge, and expanding the process of knowing.

The neurodevelopmental theory outlined in the last three chapters argues that the human brain provides the capacity for structuring experience through its essential motive controls. We differentiate the specifics of the world through adapting to the errors of our more elementary expectancies, and this involves the motive control of anxiety. We integrate the general meaning of experience through assembling the commonalities of differentiated features, and this requires organizing new, more complex expectancies under the agency of a positive (elated) arousal. With this background, we can define a concept in the context of an embodied natural philosophy. *A concept is a set of motivated expectancies based on the affordances we have experienced from specific objects and events, which lead us to expect certain affordances in future experience.*

Your concept of a *chair*, for instance, is based on the affordances provided by certain identifiable types of objects you have experienced. This constitutes a developing set of expectancies or predictions about what you will experience in the future relative to things of this kind. For example, you will typically be able to sit on chairs; they will have a seat, back, and four legs; and they will support the weight of the sitter. However, not all chairs will manifest this unique set of features. According to cognitive prototype theory (Lakoff 1987; Posner & Keele 1968; Rosch, Mervis, Gray, Johnson, & Boyes-Braem 1976), there will be a central prototype (or a small number of prototypes) of a chair specified in a cognitive model of features shared by central members of the category, but there will be other nonprototypical members of the category that share certain features while lacking others. Consequently, most of our concepts cannot be reduced to a feature list of essential properties. For example, there are barber chairs (one leg), bean-bag chairs (no fixed seat, back, or legs), electric chairs (designed to execute you), posture chairs (no back), wicker hanging chairs (no legs, suspended from above), and boatswain's chairs (no back, no legs, suspended from above). In short, *chair* forms a "radial structure" (Lakoff 1987, 83), with prototypical chairs central to the category, surrounded at various distances (where distance metaphorically represents degree of shared features) by nonprototypical members. Consequently, prototypical members of the category are specified by our high expectancy of certain affordances, with nonprototypical members having lower expectancy ratings for some of the affordances so appropriate for prototypical members.

The prototype theory of category structure reveals that our most popular and deeply rooted metaphor for concepts is seriously mistaken. What Lakoff (1987) calls the classical objectivist model of categories assumes the traditional metaphor CONCEPTS ARE CONTAINERS, which regards the members of a category as a set of individual things inside a concept container that is defined by a set of necessary and sufficient conditions. When the CONTAINER metaphor is taken too literally, the conceptual boundary is regarded as clear and rigid, an all-or-nothing affair. A particular thing is either in the category boundary or outside it. A chair is a chair is a chair, right?

Wrong. Prototype theory replaces the CONTAINER metaphor with an image metaphor of RADIAL STRUCTURE. Within this metaphor, we think of category structure as a distribution space in which the probability of an object's having certain characteristics is arrayed. Some objects will be at or close to the central prototype(s). Others will be at various distances from the prototype, though still recognized as part of the category. Others will be, in a particular context, seen as outside the category. The notion of a clear-cut category boundary with fixed internal structure gets replaced by a radial structure with fuzzy boundaries that can change as new members get added to the category. Whereas some think of *chair* as a concrete, fixed concept, we see that some degree of abstraction is required to understand even the prototypical examples of chairs.

Traditional metaphors for concepts (e.g., CONCEPTS ARE CONTAINERS and CONCEPTS ARE OBJECTS) turn concepts into quasi-objects that can be metaphorically seen, grasped, manipulated, analyzed, and constructed. There is a strong tendency to think of concepts as abstract entities with clear, definite boundaries and a fixed internal organization. We talk about having an idea (as a possession of mind), giving up or discarding an idea, getting an idea from someone, or giving them an idea. According to this metaphor, concepts are viewed as transferable valuable objects. This is a good example of Whitehead's (1925/1997) *fallacy of misplaced concreteness* in objectivist semantics, where an abstraction is treated as if it were a concrete entity.

It should be clear now that ideas or concepts are *not* fixed objects of any sort, whether concrete or abstract. First, they are not entities, and, second, they are not eternally fixed in their internal structure. Instead, concepts are patterns of expectancies about the affordances of experience (past, present, and future). To say you *have* a concept is not to say that you own or possess some abstract entity, but rather that you have developed, over time, a set of expectancies or habits (realized in activations of functional neural clusters) about the affordances you have experienced and will possibly experience in the future. As we rejected the WAREHOUSE metaphor for memory (as retrieval of stored memory-objects), likewise we must reject any objectifying meta-phors for concepts (ideas as mental entities). Concepts are not objects, but rather patterns of expectancy propagated across cortical areas, and fully rooted in our motivational systems. A concept exists in a neural network, operating a human body that is engaging its environment,

Objectivist semantics claims that we must already have a pre-given objective conceptual definition of what constitutes a particular kind (e.g., the kind *chair*) before we can identify a particular instance as a chair. To the contrary, we suggest that kind terms are probabilities about how certain individual things or events provide similar affordances in our ongoing experience. We recognize prototypical chairs because our past experiences of them have generated strong expectations about what chairs look like, how we interact bodily with them, and what we use them for. However, when we encounter new objects that share at least some features with our prototypical chair but lack others, we recalibrate our expectancies for affordances of chairs, thereby refining our concept of a chair, now manifesting radial category structure.

Instead of membership in a category being an objective once-and-for-all, all-or-nothing demarcation, our values are recognized to play a crucial role in how we understand a specific category in a specific context. A barber chair is a great chair in which to get your hair cut but a terrible chair for a dinner party. A boatswain's chair is great for transferring a sailor between two ships at sea but lousy for sitting in to read a book. Our definition of *concepts as expectancies of affordance clusters (i.e., a probability distribution of predictions about experiences)* situates concepts in their proper motivational context. Concepts are embodied, experiential, rooted in the motive con-trols, and based on our biological and cultural values.

11.2   Complexity of Knowledge Creates Complexity of the Self

In the traditional quest for certain knowledge, analytic philosophers and cognitive neuroscientists reject subjective aspects of knowing. However, by doing so they deny the scientific evidence for the adaptive basis of the mind's function. As it organizes meaning, the process of knowing links us accurately to the complexity of the world and clarifies the personal, subjective signifi-cance of that complexity. Looking at the adaptive mechanisms for motiv-ing the conceptual process, we discover that the control of neural activity in time leads to specific forms of conceptual structure. The brain's motivational controls are the mind's engines for organizing abstract—differentiated and integrated—conceptual complexity.

Through this embodied, neurophysiological analysis of concept formation, arising through the motive control of differentiation and integration, we may bring our developing theory full circle, back to the embodied sources of abstract conceptualization and reasoning that we explored descriptively in chapter 6. The collective insights of the culture are captured in our shared concrete perceptual concepts, as well as in conceptual metaphors that are manifested in language and other forms of symbolic interaction. Through these shared concepts, the embodied processes of sensing, feeling, and acting become the foundations for thinking abstractly about the regularities of the world.

The structure of mind is organized from cultural foundations to be articulated in each person through embodied structures of meaning guided by motive control. Building objective knowledge requires understanding the rich complexity of the environment. To understand objectivity, we must specify the meaning of information in the physical world, subject to inter-subjective conceptual frameworks provided by our social world. In order to fit the differentiations and integrations that define the complexity of relations among the elements of reality, there must be corresponding structural complexity in the developing mind.

The construction of complex knowledge of the world requires a parallel complexity of the self. Just as the neural networks of the brain organize all of the person's experience within the adaptively integrated architecture of the organism, the constructions of mind are formed within the entirety of the embodied self. There is complexity in the world, and knowing makes this the complexity of mind. Since self and world are co-constituted, and since mind is an emergent functional system of shared meanings, our cognition is always *in* and *of* the world, and there is no fundamental gap between mind and world. The process of knowing is therefore a process of self-definition as much as it is reality apprehension. The more complex organization of the relation of mind to world that is achieved by abstract concepts creates a more abstract and flexible construction of the self. This is fundamentally a subjective process, one that not only creates self-awareness but soon depends on it.

11.3   The Subjective Cost of Knowledge

The realization that the mind's neural networks are highly integrated, linked as we saw in chapter 7 through ordered connections with each region of neocortex regulated by its limbic base, makes it clear that learning new information must be negotiated in relation to the disruption of old learning. In the early research on integrated artificial neural networks, this became known as the *stability-plasticity dilemma*, in which the stability of existing learning is threatened by new learning (Grossberg 1980). Although there are important mechanisms of the mammalian brain to maintain stability of existing knowledge, including memory consolidation during sleep (Wei et al. 2018), these are not guaranteed to avoid the stability-plasticity dilemma, and significant new learning may disrupt the knowledge held within the mind's representational networks.

This challenge of mediating between learning and stability may be seen as a technical problem for neural network theory, but when we recognize the coherence of mind as a property of the embodied organism, it becomes an existential and psychological problem for the person. When a significant experience involves information that is meaningful for the person, the process of learning may require a revision of knowledge and of the self. Knowing in this sense is a negotiation between maintaining one's current identity and learning something new. When the personality is organized around concrete conceptual systems, this negotiation of identity and learning is implicit, such that the person whose identity is fragile or threatened may not be able to handle the personal cost of new learning (Harvey, Hunt, & Schroder 1961). However, for the person who integrates abstract thinking more fully within the personality, the process of knowing still requires change of the identity, but the flexibility of mind associated with abstract thinking allows a more fluid integration of new information within the conceptual structure of one's personal identity.

11.4   Abstraction through Self-Awareness

The transition to abstract thinking in adolescence provides an important illustration of the psychological principles that explain how the increasing complexity of mind is associated with a more flexible identity when self-awareness becomes more fully developed. Jean Piaget studied the transition from concrete to abstract thinking that occurs as children develop into early adolescence (Piaget 1936/1992). He found that adolescents become able to understand abstract concepts at the same time that they become acutely self-aware.

This coincident emergence of abstract thought and self-awareness is consistent with the idea that concepts are representations not only of the relations *in* the world, but the relations of the self *to* the world. As the process of knowing becomes more conscious, it becomes more deliberate and flexible, able to engage in hypothetical thinking. As the child develops into adolescence, she is no longer bound to the immediacy of her present concrete experience. She becomes capable of seeing situations distant in space and time that exhibit shared relations with things and events as presently experienced.

As philosopher Johann Fichte (1796/2000) explained, we become aware of ourselves largely through the process of relating to others, first understanding that others have their own perspectives, and then appreciating how we appear to them. The concept of ourselves as separate entities who have our own perspective that may differ from others and the flexibility of holding our ideas tentatively appear to emerge simultaneously by engaging in interpersonal perspective taking with family and friends. This is the fundamental process of intersubjectivity, creating an interpersonally grounded, and more veridical, self-awareness.

Arising from its egocentric childhood origins, the acute self-awareness of early adolescence emerges at the same time as the capacity for thinking abstractly and flexibly about problems. The adolescent's implicit construct of self arises from its base in childhood family and peer relations to become articulated with new complexity in adolescent peer relations. Piaget described abstract thought as *formal operations*: the capacities for forming hypotheses and making deductions from them. Whereas the young child can only think in one way, egocentrically identified with one concrete idea of what's happening, the adolescent can adopt the *as if* attitude, in which an idea is a hypothesis representing only one possible way to see things. This is a kind of freedom in which the self is no longer fused with its concrete present situation, but instead can pick up new ideas tentatively and manipulate them creatively.

This differentiation of the implicit self from the process of cognition leads to a powerful new flexibility, a kind of rebirth of identity, throwing off the egocentrism of thinking that one's personal perspective and reality are the same. Through learning to appreciate the psychological perspectives of significant others, the increasingly mature young person recognizes new relations among different things in a way that deepens and enriches meaning. This emergence of self-aware cognition is the birth of conscious knowing. It is a new order of mind in which expectancies emerge not merely from needs and motive urges but through the active process of reflective self-direction.

An example of thinking abstractly in the objective sphere, about objects in the world, is the child's learning how weights are used on a balance. The further away from the balance point that the weight is placed, the less weight is needed for balancing a given reference weight. Piaget observed that preschool children cannot understand the task of selecting the weight to achieve balance at a given position on the lever arm. In contrast, school-age children are often able to find the right weight, albeit typically through trial and error. Adolescents demonstrate the capacity for formal operations through understanding the principle of leverage (the relation between distance on the lever arm and weight), as well as the ability to test this principle through systematic manipulations. The principle is an abstract one, requiring both differentiation of the elements of the problem (weight and distance from the balance point) and also the conceptual integration that yields the insight of their invariable (and inverse) interaction.

To appreciate how self-awareness supports this sort of abstract thinking, Piaget studied the adolescent's emerging *metacognition*—the ability to think about thinking. Metacognition marks the transition out of the egocentrism that is integral to the young child's concrete thinking. Through metacognition, the adolescent can adopt the *as if* attitude of abstract thought. For example, maybe if the balancing weight is farther from the fulcrum, then the balancing weight can be lighter. Coming up with a prediction like this requires that the concept is not a fixed certainty, but rather a hypothetical that can be flexibly entertained for its merit in explaining objective evidence. *With the tolerance of uncertainty required to hold them tentatively, as hypotheses, concepts become flexible tools for investigating our experience.*

The concept is still a relation to the world, but in the presence of self-awareness, the relation to the world is now more explicitly conscious, unbound from syncretic concrete fusion with the present situation. Greater self-awareness allows metacognition because the mind is no longer fused with the present situation and engages each way of thinking about the problem tentatively—at a distance from the self. This distance from the self is achieved primarily through intersubjectivity, through experience with understanding how others see us. Because it is tentative, more distant from the self, the abstract concept can be more easily criticized. It is also inherently objective, a property of the world rather than an extension of the self. Achieving formal operations—and the self-awareness that allows it—serves as the necessary condition for critical thinking.

We interpret Piaget's observations to suggest that an increase in self-awareness is important in the development of abstract thought for the very reason that concepts are not just maps of things in the world, but also ways of relating to the world. They help us to understand what things mean for our lives. As a result, the advances in the structure of intelligence required for abstract thinking begin with the capacities for forming differentiations and integrations required for organizing complex concepts. Building on these structural foundations, metacognition allows a new level of flexible control of conceptual capacities in the process of self-regulation, leading to an expanding consciousness. Consciousness includes not only awareness of the complexity of the world, but also awareness of the knower's personal, meaningful, and subjective relation to this complexity. For the fortunate adult who gains the capacity for abstract thinking, this enhanced self-awareness includes the capacity for allowing the identity to evolve in the experience of significant new knowledge. The inherent subjective basis of knowledge requires that subjectivity is modifiable. In an important sense, this is the cost of objectivity.

A paradoxical challenge for organizing a theory of conscious knowledge is that metacognition becomes important to the process, even though the person is only vaguely conscious of being self-aware. We infer from several lines of evidence that the motive control of cognition is subjectively important—shaping the process of personal experience—even as we have only limited conscious awareness of the process. At the same time, the capacity for metacognition seems to require self-awareness, even if it remains rather limited. Understanding consciousness in modern scientific terms therefore undermines the traditional assumption that rational knowledge is assembled in the clear daylight of consciousness. Instead, we find that the process of mind is largely unconscious, and yet the ability to participate as agents of this process is essential to abstract intelligence. To become effective agents of the embodied mind, we operate—more or less competently—in the twilight of self-awareness.

Each person who develops abstract intelligence gains the capacity for an awareness of knowing that provides a higher-order relationship to the world. When bound by concrete thought, the person is fixed in a relationship to the world defined by that knowing. Thought is rigid and deterministic. In contrast, abstract thought emerges in concert with self-awareness as a direct result of the higher-order control that is possible for the conscious mind. Each thought becomes a possible way of seeing the world, not an unavoidable necessity. The effect of this higher-order control is the flexibility for organizing the complex structure of abstraction, so that creative integrations can be envisioned while remaining conscious of the differentiations necessary for accurate reflection of the complex relations within the world.

When self-organization reaches a level of complexity suitable for abstraction, we enter the realm of shared meanings that Dewey (1925/1981, ch. 7) called *mind*. Mind constitutes an interpersonal and cultural level of complex communicative interactions that makes possible abstract conceptualization and thinking. Consequently, what most people think of as their individual private mind (locked up within their distinct conscious self) is actually an emergent functional system of cognition and meaning shaped by sociocultural forces and events. As we saw in chapter 4, mind is an interpersonal process. Therefore, when we speak of individual minds or private minds, we must always remember that these are only possible within culturally shared systems of meaning and thought. What I think of as *my mind* exists only as a participation of my embodied self in culturally shared forms and systems of meaning making.

11.5   Self-Regulating the Motive Control of Conscious Intent

We have argued that the emergence of self-awareness makes abstract thought possible, giving rise to a new level in which thinking can reflect on its own operations, and can function more flexibly and hypothetically. How is this emergence of complex abstract thinking possible? A key transition in the neurodevelopmental process is the maturation of the frontal lobes in adolescence. Based on examination of the deficits in self-control observed in people with frontal lesions, psychologists have suggested that slow frontal lobe maturation may explain the immaturity of self-control in adolescents. Adolescents encounter new adult urges and interests but lack the judgment and self-control provided by a fully functioning frontal lobe.

This interpretation makes sense. At the same time, it is important to recognize the positive skills in self-regulation that mature at the same time as the frontal lobe. These self-regulatory skills, tested by newly adult urges, form the foundation for self-awareness.

The human frontal lobe is unique in its widespread neural connections to other brain regions, supporting the *executive functions*. These uniquely human capacities depend on the basic mammalian cerebral architecture, but the human frontal lobe seems to allow us a qualitatively new level of abstract self-control. These executive functions include directing attention, organizing working memory, and managing self-control. The most direct frontal connections are with the motor cortex, so that the executive functions are involved in generating and regulating action. However, the frontal lobe is also integral to the control of subcortical systems that regulate brain arousal and also bodily (autonomic) arousal in support of motivation and emotion. These subcortical systems include the brain stem and thalamic circuits regulating neural activity, and they include the limbic circuits that organize the ongoing consolidation of memory—consolidation that proceeds under motivational control (Barbas 1995). Although the frontal lobe gains its executive control through intimate collaboration with multiple networks, it is a kind of convergence zone for the adaptive, motivated control of the embodied mind (Damasio 1998). It binds together motivational control, emotions, motor control of actions, and memory consolidation.

The anterior pole of the frontal lobe (behind the forehead) has expanded more than any other cortical region in human evolution (Semendeferi, Armstrong, Schleicher, Zilles, & Van Hoesen 2001). It is therefore interesting that the frontal lobe is the last region of the brain to mature in child development, bringing a higher-order control of the massive networks of the neocortex. This occurs during adolescence with the appearance of abstract thought. By considering the anatomy and neurophysiology of human frontal networks, it may be possible to explain how expanded frontal control is important to both the structure of abstract thought and the capacity for self-awareness that goes with it.

Conscious experience requires activation of widespread cortical networks. Tononi and Edelman (1998) argue that we don't achieve consciousness with only the activation of a single functional area of the brain. Instead, consciousness arises when there are multiple cortical networks interacting together. The essential function of consciousness thus appears to be supporting a common space of working memory that allows multiple systems to coordinate a coherent process of cognition (Baars 1986). As a result, the core limbic cortices that regulate memory consolidation must provide essential support for the continuity of memory that is the necessary backdrop for conscious experience.

If there are multiple cortical networks operative together to create a moment of consciousness, where does the unity of consciousness come from? The unity of consciousness has long been a key philosophical question. David Hume worried that introspection reveals only a succession of conscious states, but without direct awareness of a self that binds together those states into a unified consciousness. Immanuel Kant replied that there must be an underlying source of unity of the self because "the *'I think'* must be able to accompany all my representations, *as mine*" (Kant 1781/1968, B131). Kant thought this unity of the self had to be a transcendent, non-bodily ego—a source of pure spontaneous organizing activity—but there is no scientific evidence for such a noumenal reality. Instead, there is the embodied self that requires the continuity of ongoing memory in time. If you cannot remember what just happened, you may appear conscious, but the real, functional continuity of your mind has been lost. Any insult to the brain that causes an actual loss of consciousness, such as a head injury or limbic seizure, also disrupts ongoing memory. With ongoing memory disrupted, as the person becomes conscious again, they are disoriented as to where, or even who, they are (Tucker & Holmes 2010).

Yet, there are specific seizures that disrupt the frontal pole specifically, and these cause a unique disruption of conscious intentionality that does *not* impair ongoing memory and that does not impair orientation to self and context. In childhood absence epilepsy (formerly called petit mal epilepsy), the child may stop what she is doing, stare into space, and be unresponsive to someone speaking to her. We know this involves a disruption of the frontal pole (as well as its thalamic control circuits) because electroencephalogram recordings show large spikes that engage the frontal pole specifically (Tucker, Brown, Luu, & Holmes 2007). Importantly, this kind of seizure does not impair ongoing memory. We can tell this because once the seizure is over, the child can immediately continue the conversation with full memory of what was being said (Niedermeyer 2003).

The implication from these clinical observations is that human consciousness is normally assembled from two components. One is the continuity of ongoing memory, which requires the continuous support of cortical traffic by dorsal and ventral limbic circuits at the core of the cerebral hemisphere; the other, conscious voluntary control, may be uniquely human, or at least it is uniquely developed in humans. Only humans have been observed to have absence seizures (Niedermeyer 1966), and these often disappear with frontal lobe maturation in adolescence. The specific aspect of consciousness that requires the uniquely human frontal pole seems to be the voluntary agency that allows deliberate control of ongoing thought and behavior (Tucker et al. 2007).

Thus, whereas the limbic motive control of ongoing memory is inherent to mammalian cognition generally, human consciousness seems to elaborate the self-regulatory circuits of limbic, thalamic, and neocortical networks in a unique way that gives a more extended neocortical control over the self-regulation of arousal, attention, and memory. In neuropsychological terms, this can be called *representation of the regulatory function* (Tucker & Luu 2012). Whereas motive self-regulation in most mammals is direct and reflexive, with only implicit mediation by representation (concepts), humans develop a more deliberate, effortful, and we could say thoughtful capacity for self-control. On the basis of several lines of evidence, we can attribute this to our uniquely enlarged frontal pole of the neocortex.

As the frontal lobes mature in adolescence, the capacity for deliberate conscious control of cognition seems to mature as well, leading to a new competence in self-monitoring that allows metacognition. This ability to reflect more directly upon the process of cognition is important to adopting the flexible as-if attitude that allows the abstract thinker to entertain multiple perspectives on an issue. Whereas both dorsal and ventral frontolimbic systems provide adaptive control of ongoing memory—and the continuity of experience—the frontopolar networks seem to provide higher-order awareness and monitoring of the cognitive process itself. With this new capacity, it should not be surprising that the adolescent feels that he is ready to take control of his life.

In this sense, the maturation of consciousness in adolescence appears to be an essential basis for the development of self-consciousness. This self-consciousness provides the executive, metacognitive control of the structure of cognition required for abstract thinking. The developing capacity for conscious self-regulation contributes a kind of supervisory process, a level of mind in which each act of cognition is not the sole process of experience, as in concrete thought, but is one component of experience, a way of expecting the world that can be entertained and kept or discarded without disrupting the continuity of awareness.

Recent scientific accounts of consciousness have emphasized the functional role it plays in human cognition, providing the integrative workspace for mental operations to share common representations (Dehaene & Changeux 2011; Edelman 1987; Tononi & Edelman 1998; Tononi & Koch 2008). The functional roles of frontal networks appear to allow a higher-order, deliberate capacity for both monitoring and managing expectancies and their interaction with error correction across widespread cortical regions. The neural architecture we have been examining appears to constitute the neurophysiology that makes metacognition possible. Each network's traffic, across limbic, heteromodal, and sensory/motor cortices, mediates prediction and error correction. The frontopolar networks and their limbic and thalamic counterparts may provide a higher-order representation in humans, able to support a supervisory—conscious—integration of this expectant process. The tuning of uncertainty, coupled with the associated motive states of anxiety and elation, is an essential implicit basis for the frontal executive function. Understanding the metacognitive control of uncertainty offers insight into the subjective process of explicit consciousness, with the self-awareness required for achieving abstract thought.

11.6 Personality and the Adaptive Control of the Executive Functions

Knowing has traditionally been conceived as an intellectual process, devoid of emotions and feelings. That cannot be the case, given the fundamental role of motive controls in the process. Are the constitutive values and emotions unconscious, or is there a subjective feeling of knowing? Important clues come from the unique forms of control provided by dorsal and ventral limbic circuits; these are dual ways the maturing frontal lobe engages limbic motive controls.

Subjectively, the sense of agency and initiative integral to consciousness is supported by the dorsal frontolimbic circuits (Tucker & Holmes 2010). We might describe the conscious subjective quality of this active, feedforward mode of control as *intention*—the feeling of launching an action. This is the feeling we sometimes have of projecting our energies toward realization of a valued goal. Perhaps this is what James meant when he described the feeling of furtherances.

The fact that dorsal frontal lesions lead to pseudodepression implies that, normally, a positive, impulsive motive bias provides an integral influence on expansive thinking and conceptual integration. This is a feeling of agency, of being in control and confident in what's happening. The pathological distortions of cognition are instructive as well. In clinical mania, the expansive grandiosity associated with unbridled positive affect represents the exaggerated form of positive motive control on integrative cognitive structure.

The other important feeling is the anxiety associated with activity in the amygdala and ventral frontal circuitry that typically leads to both focused perception and an analytic conceptual style (Derryberry & Tucker 1991). Although this form of control may be exaggerated, such as in obsessive-compulsive disorder, it is normally a productive, yet fully unconscious, capacity for conceptual differentiation. We might call this ventral limbic mode a process of *attention*, with its own subjective quality of uncertainty, the need for constraint, and an emphasis on feedback from the world. Phenomenologically, this ventral process is felt as a sense of anxiety, hesitation, and doubt, with the feeling that we can't quite go forward under our current conditions of indeterminacy. This feeling of frustration and resistance is likely what James meant by felt hindrances.

The motive controls on memory and cognition are then realized to be subjectively significant. We experience them as the subtle moods of the day. In our simplified theory we summarize the core moods as anxiety and elation, but the actual moods of experience are more varied and nuanced, and are of course woven together in much greater complexity. What is important to realize is that as these mood states and their inherent motive controls are engaged, they simultaneously shape conscious experience, the organization of concepts, and acts of reflective knowing.

As we have seen, the control of thought is by no means conscious in a simple sense. However, it does appear that the development of self-awareness gives us some perspective distance on our thought processes, some capacity to be aware of thinking (metacognition). As a result, ideas are entertained as possible ways of knowing rather than absolute, concrete assertions of what world and self are like. Through ongoing subjective experience with intentions, we learn to rely on the furtherances of thought. As we encounter the challenges of attending and adapting to the constraints of reality, we learn to deal with the hindrances.

An important theoretical implication from this line of reasoning is that knowing involves inherent biases toward managing uncertainty that are integral to organizing both abstract cognitive structure and the process of awareness. These motive biases generate qualities of consciousness, even if their operation is largely unconscious. We are suggesting that the structural influences—and likely the biases in handling uncertainty—are basic mammalian modes of self-regulation. However, with the evolution of the elaborate networks of the human frontal pole, recruiting both thalamic and brain-stem controls to tune the activity of widespread cerebral networks, there seems to have evolved a new more flexible capacity for deliberate conscious control of the process of knowing. *In other words, humans, with their more highly developed frontolimbic capacities, have recruited their mammalian motive control systems, which originally served perception and action, for higher-level processes of conceptualization, reasoning, and self-awareness. Hence, all acts of conceptualization and knowing are made possible and shaped by our most basic motivational processes that ultimately serve our survival and flourishing.*

11.7   The Intersubjective Construction of Self and World

Through the young person's experience with intersubjectivity, self-awareness develops within the context of the perspectives of others. Even as we look to the embodied basis of mind in the neurobiological mechanisms of motive controls—of anxious differentiation versus optimistic integration—we can appreciate that these mechanisms evolved because learning is social. Self-awareness allows the complex, metacognitive form of knowing because the self is constructed within the interpersonal context.

Knowing is thus not only embodied. It is embedded within the experience, institutions, and shared practices of the culture. It is an intersubjective process, not just an individual achievement. The classical insight into the social construction of mind came from Fichte's (1796/2000) dialectical theory. Fichte's dialectics proposed that the antithesis arises in opposition to the thesis, and helps to define a new synthesis. Particularly as elaborated by Georg Friedrich Wilhelm Hegel, there is a fundamental thesis-antithesis-synthesis structure of ongoing inquiry that has been described also by thinkers as diverse as Karl Marx, Sigmund Freud, and Thomas Kuhn.

Fichte, in turn, was a careful student of Immanuel Kant's philosophy of mind, in which reality is not simply known, but rather is co-constituted by mind and the evidence of the world (Kant 1781/1968). Kant saw that our world as experienced is the result of sense perceptions given to us through the sculpting activity of our shared forms of spatial and temporal intuition, which are then cognized via concepts that are either innately given (i.e., the pure concepts of understanding) or learned (i.e., empirical concepts arising from recurring encounters with objects and events in our world). Because the structure of mind is integral to the process of knowing, we know the world in part through creating the world as we know it. We can have genuine knowledge of our world because we have a major hand in shaping that world and giving it meaning. Our selfhood arises in and through its synthesizing of conditions in our world.

Starting from Kant's perspective on the co-constitution of self and world, Fichte added what seems to be a practical and even prosaic twist. We gain knowledge through our experiences with other people via shared meanings. Knowing is organized in the process of social interaction, a process that when fully developed within the complex mind becomes one of understanding and sharing the perspective of the other. In modern psychological terms, we describe this as intersubjectivity.

In Fichte's reasoning, the result of intersubjectivity is a new self-awareness of subjectivity itself. As we understand that another person has a unique subjective viewpoint, different from the perspective we have, our own subjective consciousness becomes more fully appreciated as a separate process, no longer concretely fused with the childlike egocentric apperception of the world.

Fichte's intersubjectivity anticipated Piaget's *as-if* reasoning, as it arises specifically in the context of interpersonal experience. This is a powerful context because we have the awareness of another mind (or other minds) to guide the awareness of, and distance from, our own subjectivity. The increased flexibility and tentativeness of ideas—as opposed to concrete fusion with an egocentric view—then becomes the essential basis for abstract intelligence.

Hegel, Marx, and Freud are not the only thinkers whose views were profoundly influenced by Fichte (Marcuse 1941). Intersubjectivity has become an important topic for research into the child's development of insight into social relations (Jackson, Meltzoff, & Decety 2005). As we attempted to formulate the adaptive basis for self-regulating the process of knowing, we recognized that the adaptive controls operate strongly in the intersubjective as well as the subjective arenas.

In a generic account of the feeling of knowing, the initial impulse for conceptualization arises with the agency of certainty, a kind of extraversion in which self and world are fused. In the interpersonal arena, the psychological process is one of identifying with the new idea and, of course, its social source. This is the *thesis* of accepted knowledge, motivated by the elation of extraversion. It is the foundation for the *paradigm*, the knowledge that unites a scientific, political, or other intellectual community. Consciousness at this stage is egocentric, syncretic, and concrete in that the self and the idea are undifferentiated. This is the identification stage of knowing, a kind of infatuation, whether with a person, a theory, or a political or religious orientation. It involves the confidence of shared meanings, the joy of sameness, and the satisfaction of being understood by others and understanding them in return.

The next stage of development is one of individuation, or partial separation, from the identification. The perspective of intersubjectivity is important because the adaptive control of individuation engages anxiety and hostility as the motive engines of the *antithesis*. The critical attitude that results is almost as concrete as the identification with the idea at the stage of the thesis. When motive controls are strong, the negativism breaks the self from the intersubjective perspective. Without further development, abstraction suffers here as well.

When motive controls operate in service of the process of knowing, and do not drive it inappropriately, as in the personality disorders, then the dialectical process allows both identification and individuation to achieve a higher-order integration, the *synthesis*. In the analysis of conceptual structure presented in these pages, the thesis is a kind of syncretic whole, the collective expectancy of the shared paradigm. The antithesis is the critical thinking and error correction that is achieved through individuation of personal perspective from the collective assumptions. The synthesis is then a new integration that has the structure of an abstract concept because it not only spans the scope of the original holism, but also has incorporated the differentiations achieved through the critical analysis of the antithesis. The new paradigm that results then achieves both greater complexity and a greater fidelity to the complexity of reality.

Thus, when the interpersonal context is considered, the process of knowing is not only a biological one but also a cultural one in which intelligence is never a separate property but is achieved through social communication.

Objectivity is then achieved through intersubjectivity—the capacity for abstract insight into the mind of another that shows that reality is not fixed by our egocentric (and ethnocentric) perspective, but rather knowable in an abstract sense by another person. The process of mind is then both embodied and cultural—a process of maintaining an identity through knowing, and also entertaining the subjective worlds of others, as we continually organize the relations of mind to participate in the events of the world.

11.8   The Process of Knowing in Science

The personal, subjective process of mind must be engaged in each achievement of objective knowledge, including science. What we call the scientific method has been a powerful way of advancing human knowledge by emphasizing objective evidence that is gathered and tested in relation to carefully formulated hypotheses (Whitehead 1933). Various empirical sciences, each with its own distinctive methods of inquiry, share a general pattern of observation of phenomena, hypothesis formation, modeling, controlled testing, and readjustment of the initial hypothesis as required by discrepant evidence. It was remarkable to Piaget to observe the development of formal operations in adolescence and to recognize the natural and spontaneous emergence of the capacities for explicit hypotheses and deductive reasoning that form the basis of scientific inquiry. This was Piaget's (1971) contribution to what he called *genetic epistemology*, the study of biological development to describe the origins of knowledge.

The hypothetico-deductive process of classical scientific method is almost literally an externalization of the neural process of predictive expectancy and error correction that is being discovered in modern cognitive neuroscience. This new discovery recalls Dewey's (1925/1981) assertion that science—empirical methodology often mediated by multiple technologies—is an elaboration and extension of modes of inquiry intrinsic to our mundane acts of knowing. Dewey noted the continuity between our daily acts of problem solving and sophisticated scientific inquiry.

Even as they adopt prediction and testing evidence as the valued methodological model of their disciplines, scientists have little training in either self-awareness or subjective reflection. This same bias is passed on to their students, who are typically indoctrinated with the preferred illusory ideal of pristine objectivity. As a result, the scientific method remains as an externalized discipline of action. This discipline functions as a kind of unconscious ritual of the specific mechanisms of information processing that has evolved within the mammalian cortex, instead of becoming part of the conscious tools of the scientifically self-aware.

How could we approach the process of scientific inquiry with a more explicit self-awareness? The theoretical analysis we have provided sees the process of knowing as an active and motivated process in which the mind gains a personal relationship with the world. Because concepts are relations of self to the world, gaining knowledge is a process of relationship building. What becomes abstract, then, is not only objective concepts of aspects of the world, but also our personal, subjective relationship to the world. To be most effective, scientists should then be trained to become self-aware. The implication is that a science without self-awareness is a degraded form of knowing, a kind of concreteness of the unaware.

Recognizing that the neuropsychological mechanisms for the frontal lobe's executive control of cognition play a key role in the maintenance of consciousness, it should become clear that motive control of intelligence is a subjective process. It is not conscious in the explicit sense of knowing exactly what's happening, but in the sense that we recognize that our ideas (concepts) are not reality—fixed, finished, and whole—but rather tentative expectancies for what it might be. For the scientist, these are hypotheses—methods for instituting and shaping further inquiry.

The conclusion from our biological analysis of knowing is that the motive states of anxiety and elation are subjective guides to the ongoing conceptual learning process. We think by feeling. Certainly, the validity of concepts must be established objectively (i.e., via testing against experience). But even the mechanism for specifying the necessary differentiations for accurate representation of reality, and thereby handling the errors when expectancies are wrong, turns out to be an adaptive mechanism of motive control. Under conditions of threat, anxiety augments the perceived uncertainty of concepts, such that our trust in the knowledge is tenuous, and we search for evidence for objective validation, even though such validation can never be final, complete, or absolute.

The subjective basis for knowing—experienced as confidence in one's cognitive processes—is often evident in cases where we intentionally generate novel integrations. This generative process of mind seems to depend on both the confidence of personal agency and the expectancy for success.

At the same time, there are distortions of subjectivity, and these are well known to science. There is clear evidence of the exaggerations of subjective intentionality waxing and waning in the mood disorders of mania and depression. There are well-documented subjective distortions of exaggerated certainty in the hypomanic and histrionic personalities, as well as exaggerated uncertainty in anxious and obsessive-compulsive personalities. Scientists are not immune to such distortions, which show up both in overconfidence in their research and theories, on the one hand, and in extreme skepticism and hypercriticism, on the other. The theoretical implication seems to be that each instance of knowing requires adequate self-regulation through the subjective process of engaging the motive controls that tune not only expectancy and uncertainty, but also the associated integrations and differentiations of conceptual complexity.

11.9 The Dialectical Progress of Scientific Knowledge

Knowing can thus be understood as a dialectical process in which a relationship is formed with the phenomena of the world we are trying to understand. This requires a balancing of critical appraisal—through the differentiations of anxiety—with acceptance and identification—through the integrations of personal agency. Because human intelligence develops in an interpersonal, cultural context, the dialectical balance of criticism and acceptance is a highly personal process, with its subjective control mediated optimally by the achievement of insight, but carried out using the templates of knowledge formed by each thinker's interpersonal developmental history and by shared cultural modes of inquiry (Kohut 1978). Consequently, what we count as knowledge will depend on the values and assumptions of the knowledge communities in which we participate.

In the humanities, the historical, cultural, and personal context for knowledge is widely appreciated. In the sciences, too often the typical assumption is that the hypothetico-deductive process (or some other mode of inquiry) is a logical, objective process, untainted by subjective bias. In actuality, the historical study of science shows that the process of knowledge growth is not only highly subjective but often hotly political. The scientist and historian Thomas Kuhn (1962) studied many cases of advances in scientific knowledge. He found that what he called *normal science*, which makes up most of what scientists do, is carried out within the confines of large-scale frameworks he called *paradigms*. A paradigm is not just a theory, but also includes a scientific discipline's values, goals, methods, technologies, institutions, and practices. However, every scientific paradigm is always confronted with anomalies and discrepant evidence that cannot be immediately reconciled with the theories and practices definitive of the paradigm. Therefore, these tensions can occasionally give rise to scientific revolutions, complete upheavals in understanding, which occur only as the established order of previously accepted theory and practices—the paradigm—is overthrown and replaced by a new governing paradigm.

Since every paradigm-situated scientific theory is based on decisions about what the relevant phenomena to be explained are, what values and which methods of inquiry are best, and what constitutes a good explanation, there is no God's-eye perspective from which to select *the* one true paradigm or *the* one true method. Until a new paradigm arises, we are condemned to work within our preferred paradigm, and our best critical attitude under such conditions is one that is always mindful of the partiality, perspective-dependent, and selective character of our basic assumptions and values (within a paradigm), and always aware of the anomalies that plague even our best theories. That is why the classical pragmatist philosophers argued for a pluralistic, fallibilistic approach to scientific method and to inquiry in general (Dewey 1925/1981; James 1909/2002; Peirce 1931).

Moreover, as philosopher of science William Bechtel (2008) argues, the hypothetico-deductive method is not the only method utilized by the natural and social sciences. Certainly, many scientists use a hypothetico-deductive approach to find universal causal laws (so-called covering laws) under which particular phenomena can be subsumed and thereby explained. However, Bechtel points out that many scientific pursuits, especially in the cognitive sciences, employ a method of explanation which articulates cognitive *models* of observed phenomena. The "test" of a model is then its ability to predict and/or explain what the particular scientific discipline regards as the relevant phenomena. An example of this in the present book is our formulation of neurobiological models of the fundamental motivational control processes upon which we have based our explanation of cognition and learning. We have also referred to artificial (though brain-based) neural network models to explain conceptual structure and certain observed behaviors. In both cases, we project expectations about future observations (a form of hypothesis construction) based on the models of brain anatomy and functional cognitive processes we have articulated, and then explore whether our models can account for certain observed phenomena (a form of testing). But there is no need in this process for alleged universal causal laws.

With the principles of embodied knowing developed from advances in modern neuropsychology, we recognize the key elements of the mental activity of scientists as they create scientific revolutions. The established paradigm, the way of doing normal science that nearly everyone accepts at a given point in time, is a meaningful conceptual system, providing order for the relation of minds to the field of study and to everyday experience. It is not just an objective theory to be evaluated for its merits, but also a way of life, defining meaning and identity for those who have made their careers in this way of thinking, engaging in the practices and values that define their favored paradigm.

In this context, a scientific revolution redefines the world for all of us. The classical example is the Copernican revolution, through which we understood that the earth is not the center of the heavens. Assuming that the earth is the center of things is the natural perspective of egocentrism, the perspective that, when shared, constitutes ethnocentrism. At a sufficient level of generality, we call it anthropocentrism. Giving up this view—this shared thesis of egocentrism—has been the continuing challenge to human identity posed by each advance in knowledge. This is why a significant advance in science often requires a scientific revolution—sometimes small, sometimes of major proportions—as necessary to overcome the massive inertia of the normal science spawned by the dominant paradigm. In this sense, we can recognize that the paradigm has emotional as well as practical value. It is the shared prediction of reality that gives the sense of meaning that arises from belonging to the field and to the culture.

In the specific fields of science, those scientists who discover the errors of the paradigmatic view—because of anomalous phenomena and new forms of evidence—must take on the task of challenging the paradigm before anyone can be convinced to give it up in order to consider new alternatives. Their challenge constitutes their *antithesis* to the dominant paradigm (as *thesis*). With insight into the neurophysiological process of error handling, we can recognize that this critical stage of evaluating an accepted prediction is highly motivated by the specific ventral limbic control systems that allow the perception of differentiations in the world. The emerging paradigm presents itself as a new *synthesis* that claims to be capable of unifying many components of the prior paradigm and then better explaining phenomena that were recalcitrant to explanation in the replaced paradigm.

Most scientists are untrained in subjective insight into their own mental process. They mostly keep attention focused on the techniques they have learned for evaluating "objective" evidence and testing their hypotheses in light of that evidence. But once we appreciate the neural mechanisms of criticism and error handling, as well as the functional role of valued predictions as guides to meaning, we can interpret the motive controls underlying the cultural context of intellectual advances in scientific revolutions more explicitly.

First of all, the process of knowing in this context is indeed highly meaningful. While it reigns supreme, the paradigm is a kind of self-defining certainty. All of us who work within a shared paradigm also share its value in explaining and predicting the events of the world. We simultaneously share its value in defining ourselves as knowers of this common world. Even to recognize the errors of the paradigm is threatening, amounting to an acknowledgement of a profound uncertainty in the known world. If the earth is not the center of the heavens, then where are we? If we are the products of evolutionary development, then what is left to distinguish us from the animals? If the earth is indeed warming because of our actions, who are we that are destroying our home?

Moreover, the process of knowing is ultimately subjective—the relation of a subject to its world—and the flexibility to engage it in alternative ways requires metacognition, a kind of self-awareness. These facts set up what may be the central paradox of knowing. As we have seen, the capacity for abstract metacognition requires an advanced level of self-awareness, and yet the process of knowing remains largely unconscious, working its marvels without our reflective awareness and without our explicit control. The assumption of the established paradigm is a kind of concrete thinking, a shared egocentrism that must be given up in order to recognize the evidence that doesn't fit. The fact that the new evidence implies a different order of reality requires the flexibility to imagine a different order. Because abstract cognition emerges with metacognitive skill, the essential quality of intelligence to contribute to a scientific revolution may not be just the refined knowledge of the field, but an awareness of mind that allows entertaining different states of mind, different *as-if* interpretations of the evidence.

We can imagine the transformation of knowing that could occur if each student were trained with the principles of conscious control of the adaptive mechanisms of the conceptual process. The scientific understanding of human intelligence—the emerging natural philosophy of mind—is now providing a basis for approaching the subjective process of knowing through explicit and verifiable principles. Although we struggle to outline these principles in rough form here, it should be clear that they reflect an active, subjective process of relating self to the world. This relating shifts between identification with a way of knowing—the paradigm of knowledge in any field—and a critical rejection of that way of knowing based on a new recognition of the significant errors in the paradigm. As a result, the subjective process of science can be described as a dialectical process, one in which the creative generation of ideas must be alternated with critical thinking in order to establish alternative predictions that are not subject to the same errors of expectation and prediction spawned by the previous paradigm.

11.10 The Subjective Foundations of Objective Knowledge

The results of our study of the biological foundations of the embodied mind thus point to the basis of all knowledge in the motive control of experience. For us humans, the motive control of thought must be studied in relation to the mammalian architecture of concepts, building from the limbic motive base to reach out to the neocortical representations of sensory and motor engagement with the world. Through our unique neuroanatomy, we achieve highly flexible forms of self-control, and these forms allow the emergence from egocentrism to entertain objective knowledge. Yet, the mind's process is invariably based in the biology of the organism, which for us highly social creatures only develops within the social context.

In this way, a natural philosophy maintains what Whitehead (1929/2010) called the *philosophy of the organism*, a pragmatic account of the whole person. Instead of idealized, purely logical or linguistic constructions, mind shapes personality, with its intrinsic mechanisms reflecting the self-regulation of the personality. Through such self-regulation, we may become enthusiastic or impulsive on occasion, and this is not a separate social characteristic but an integral reflection of the mind's self-control that is essential to both subjective experience and the capacity for objectivity. Similarly, objectivity relies continually on the light of critical reason, and this too emerges as an expression of the personality and its entire developmental history. The implications for a natural philosophy of the organism are clear. The basis for knowledge is to be found not in some rational ego with its pristine logic, but rather in the rich experience of the whole person, emerging in each life from the remarkable evolved architecture of the human brain within its cultural context.

12 Outline for a Contemporary Natural Philosophy of Mind and Knowing

We are arguing for a revitalized natural philosophy, rooted in and enriched by twenty-first-century science. Since our focus is human cognition and knowing, we have drawn primarily from the mind sciences, broadly construed to include biology, neuroscience, developmental psychology, computational neural modeling, cognitive linguistics, and more. Prior to the emergence of distinct sciences in their more modern forms (e.g., biology, psychology, physics, chemistry) in the eighteenth and nineteenth centuries, natural philosophy simply meant the empirical study of nature, where nature included the inanimate, animate, and human worlds. It wasn't necessarily assumed that there could be one method of inquiry appropriate for all aspects of nature, but there was confidence that all natural phenomena could be studied empirically. In the twenty-first century, this conviction manifests as the claim that the various empirical methods (of hypothesis formation, modeling, and testing) can, when taken together, go a long way toward explaining the full range of natural phenomena. Moreover, because culture is a natural process, this empirical approach pertains equally to the understanding of cultural phenomena.

A contemporary natural philosophy is neither scientistic nor overly reductionist. That is, it would not claim that everything can be explained in a narrow scientific account within a single methodological framework or paradigm. Instead, it claims that multiple sciences, working together in dialogue with philosophy, can give us a nonreductionist understanding of our physical/social worlds.

The eighteenth-century dichotomy between the natural sciences (giving causal explanations of physical phenomena) and the human sciences (giving interpretive accounts of human experience, meaning, and behavior) needs to be abandoned or at least substantially reconfigured. That dualistic distinction was predicated on the assumption that causal explanations of human behavior provide only deterministic accounts and so are intrinsically unable to explain the experiences of humans possessed of radical freedom. This led to the supposition of the two cultures: the causal explanatory methods of the sciences contrasted with the interpretive methods distinctive of the humanities. To the extent that this dichotomy has any validity, it rests on the way that different methods or approaches focus on different aspects of one continuous process of organism-environment interactions. The different approaches are useful insofar as they reveal distinctive characteristics of emergent levels of functional organization.

The central idea is that the methods of different sciences, the humanities, and the arts, combined with a critical assessment of the fundamental assumptions of each method and coupled with an exploration of how those different approaches hang together, is all we have, and all we need, to give an adequate account of human behavior. According to this conception, a natural philosophy of mind would account for experience, cognition, meaning, action, and values as emergent functions of human organisms that have developed from our mammalian roots. It would recognize the central role of neural networks and motive controls in these cognitive/affective processes. Correspondingly, a natural philosophy of knowing would treat knowing as learning the meanings of things and events, and putting that meaning to effective use in managing our lives.

As we conceive it, a natural philosophy of mind draws on (1) cybernetic theories of control mechanisms of cognition and action, (2) a new scientific understanding of the role of sensory-motor (bodily) structures and processes in conceptualization and reasoning, and a recognition of (3) the crucial role of feelings and emotions in cognition, (4) how our biologically and culturally based values shape our thinking, and (5) how our brains and bodies make all of this amazing thought and action possible. We have argued that these and related developments in recent biologically based cognitive science have opened the way for a naturalistic theory of mind and knowing that is both supported by objective evidence and subjectively meaningful for our lives.

12.1   A Scientific Epistemology

As we emphasized at the beginning of this book, modern philosophy has been obsessively concerned with epistemology, the sources and limits of knowledge, and we have explored and criticized this concern as we construct a natural philosophy of mind and knowing. What is clear now is that the study of knowledge requires an explicit scientific basis. The traditional approach to knowledge, reflected in twentieth-century analytic philosophy and first-generation cognitive science, was an affirmation that knowledge is constituted of what are essentially linguistic mechanisms, accessible to conscious reflection by verbally facile academics. Instead, what we find from a broader scientific analysis is that language is only one cognitive capacity of the human mind, and it is in turn dependent on more basic neural mechanisms of memory consolidation, meaning, and concept formation that are representative of mammalian cognition generally. Furthermore, the insights from studying neural mechanisms are now convergent with those from studying artificial neural networks, which capture information in patterns of connections in much the same way that brains do.

The mind is thus available for examination, not just through reflecting on conscious processes but through scientific research on its constituent bodily mechanisms, operating mostly beneath the level of conscious awareness. This emphasis on computational neurobiology in no way detracts from the cultural basis of mind and knowledge. Indeed, it deepens our understanding of how social and cultural factors shape mind and knowing. The mammalian brain has evolved in the context of socialization, whether in the juvenile imitation of simple mammals or the complex and ongoing cultural development of humans. For each of the available levels of analysis, we now have explicit tools for research that allow epistemology to proceed as an empirical science, generating convergent evidence from multiple disciplines, thereby giving rise to a natural philosophy of mind.

What may not be obvious from recognizing this convergence is the new humanism that results from understanding ourselves in this explicit evolutionary and developmental way. What we discover is that knowing is a full-body contact sport involving energetic contact with our physical, interpersonal, and cultural environments. This embodied conception of knowing has profound implications for understanding who we are.

12.2   Implications for the Nature of Mind and Knowing

We will close, therefore, with a summary of the new humanism that is implied by a twenty-first-century natural philosophy. We start with key tenets of our scientifically informed account of mind, and then we set out the implications of this view of meaning, cognition, and selfhood for what it means to know.

1. Mind develops in context, through organism-environment interaction.

No living animal can survive without an at least partially supportive environment. *Everything* we experience, think, and do ultimately depends on how we engage our surroundings through perception and action. Mind, meaning, and our capacities for thinking and acting emerge from the increasing complexity of our bodies and brains and their transactions with our complex surroundings. It follows that no account of these cognitive processes can be adequate unless it recognizes the joint contribution of the evolved capacities of the organism and its interactions with the functional properties (affordances) of its environment.

2. There is a continuity between the structures and processes of mammalian learning and those of higher-level human knowing.

By emphasizing new insights into the biological foundations of human cognition, we have suggested that a naturalistic philosophy can not only appreciate how mind is profoundly embodied, but also how it is continuous with the natural order of things. There are two basic senses of continuity here. The first is that of mind and body. Instead of two ontologically different realities, body and mind are aspects or dimensions of one unified and continuous process of natural events generated through the interaction of an organism (here, a human body) with its environment, from which it can never be separated, so long as it lives. There is no need to bring in fundamentally different nonnatural or transcendent entities, causes, or values to explain human functioning. Mind is an emergent functional plane of natural events made possible by our ability to have shared meanings with other people through participation in language and other forms of symbolic interaction.

The second sense of continuity is between humans and other animals. Higher cognitive functions develop through recruitment of lower mammalian capacities for perception, movement, and feeling that are changed and expanded as they operate in more complex organisms engaging more complex environments. For example, we argued that much of what we need to understand the principles of human concept formation and cognitive organization can be found in the basic mammalian neural architecture. Within this architecture, the conceptual process is organized in the form of the representation of context and significant objects, regulated by the motive controls of impulse and constraint, forming affordance expectancies, and managing reality constraints. These adaptive mechanisms allow mammalian brains to achieve integrations and differentiations, by which we match organismic cognitive structure (realized in the brain and body) to the complexity in the world.

The qualitative advances in human intelligence appear to be achieved as we build not only on the implicit, largely unconscious foundations of the mammalian neural architecture, but on the frontolimbic networks that develop over the human life span and give rise to the metacognition resulting from self-awareness. The capacity for self-awareness enables the subjective perspective that is essential to the capacity for fully abstract processes of human knowing. Consequently, there is a fundamental continuity between our mammalian sensory and motor capacities and our capacities for abstract thought.

With the capacity for metacognition, we become able to hold our thoughts in a more tentative, hypothetical form. We are freed somewhat from the egocentric immersion in the present situation that is characteristic of cognition where there is minimal or no self-awareness. The perspective of the other that is gained in the process of intersubjective experience provides a natural basis for objectivity, arising in the interpersonal process of learning through relating. In these ways, the personal development of complexity in mind entails the corresponding development of abstract complexity in thought and feelings.

The extensions of these basic mammalian processes in human intelligence are significant, and give rise to new emergent functions, but the basic adaptive mechanisms have been conserved through the evolution of the entire vertebrate line. Humans achieve abstract concepts only through combining the differentiations of objects formed through ventral limbic focused attention with the integration of contexts organized through expansive dorsal limbic expectancies. Within the rich complexity of human society, the experience of each person becomes highly refined as it gets tuned up within that cultural setting. Thus, by studying the relevant scientific principles, we learn to trace the roots of our cognitive abilities back to our mammalian systems for perception and action, and we recognize that all our cultural achievements ultimately rely on and work through evolved capacities for perception, feelings, and movement.

3. Our deepest drives are rooted in maintenance of the conditions for life and its enhancement.

The primary value for any living thing is the establishment, preservation, and recovery of the homeostatic and allostatic conditions of the internal milieu that are necessary for life maintenance and well-being. The deepest visceral meaning of any experience concerns its value as a means to the homeostasis required to sustain human life and to the allostasis required to enhance its quality. Whatever values an object or event might have for our lives, we never escape this need to preserve and manage conditions that support a dynamic equilibrium within our bounded organic bodies in their ongoing transactions with their surroundings.

4. All knowledge is motivated and shaped by values.

Meaning and knowing are *motivated* processes, tied to the values of the organism. Knowledge is meaningful to us insofar as it bears on our visceral homeostatic needs and helps us function allostatically in our physical and social world. The connection to these needs may be immediate and obvious, such as when learning that a certain plant is poisonous is directly relevant to our survival. Or it may be indirect and somewhat less determinate, as when we are solving a mathematical problem. Although there may be no immediate existential payoff for studying mathematics, it has indirectly had profound implications for our ability to understand, explain, and to a certain degree control natural processes. In either case, our cognition always depends on motivational controls that drive adaptive engagement with our world.

5. The vast majority of our cognition happens beneath the level of conscious awareness.

All of our most important life-sustaining bodily functions (e.g., respiration, digestion, maintaining pH balance, controlling body temperature, immune responses) operate automatically and unconsciously, which is a good thing, because if we had to monitor and regulate all our life processes reflectively and consciously, we wouldn't last long.

Emerging from this visceral base, our processes of meaning making, conceptualization, and reasoning also operate mostly beneath the level of our conscious awareness and control. Therefore, no method of inquiry based solely on introspection and self-awareness can ever adequately grasp the complex processes (neural, chemical, and cultural) that constitute our cognitive functioning. To explore the hidden depths of the cognitive unconscious as it generates meaning and thought, we especially need the biologically based sciences of mind (e.g., biology, physiology, cognitive neuroscience, neural network theories, and cognitive linguistics). Yet, because much of human meaning and thought is organized through culture, we must also appreciate the collective cognitive unconscious, the implicit understandings and biases that we share with the citizens of our increasingly global culture.

6. *Evolution accounts for our brain architecture (anatomy), and then our developmental (epigenetic) experience tunes up our individual self-identity and habits of thought and action.*

Our human brains and bodies are the result of a long evolutionary history that supplies the functional brain regions and bodily structures that support all our cognitive activities. This evolutionary process is still under way, so there is no fixed human essence, even though there are relatively stable capacities for perception, action, feeling, and other dimensions of cognition. We are beings in process.

In constructing our neural architecture, evolutionary processes have established neural connections between various functional regions of the brain that provide a neuroanatomical blueprint for mind. Building upon this uniquely human elaboration of the vertebrate genetic plan, our thoughts, feelings, memories, and actions are then the result of our individual developmental history, from womb to tomb. This epigenetic development consists of the tuning up of networks of neural connections established through our ongoing experiences in the world. Meaning and knowledge resides in the *connectivity* of functional neural clusters, and the patterns of connectivity are shaped by deep motivational processes in our brains.

7. *Human brains evolved structures responsible for our sensory and motor interface with our world, but also structures that process the visceral significance of experiences for our well-being.*

Some neural patterns of connectivity among various brain areas make it possible for us to sample aspects of our environment (such as objects, their properties, and their relations to other things) through various perceptual modalities. These rely heavily on specialized neocortical architectures, but also involve limbic areas responsible, among other things, for motivation and feelings. Other neural pathways process the visceral import of our bodily states—their meaning for our survival and well-being. The meaning of anything we experience is consequently not only its sensory and motor content, but also its relevance to our visceral monitoring of our bodily welfare. Meaningful knowing thereby blends together objective evidence from the environment and subjective relevance to our well-being.

8. *Knowing consists of building and projecting our expectancies for what the world should be like, followed by testing these assumptions against actual experiences as they unfold for us.*

The traditional view held that knowing begins with perceptual sensory inputs that are synthesized and evaluated at higher levels of cognitive processing, which then initiate actions in the world. The scientifically correct view appears to be just the opposite of this. We are bundles of habits—expectancies for future experience that have been shaped by our prior experiences. We encounter every new situation with mostly unconscious predictions about what affordances our experience will reveal to us. When those expectations are confirmed by forthcoming experiences, we move forward in action more or less automatically and effortlessly, with little or no cognitive awareness or feeling. If there is any feeling, it is mild elation, or what James called a sense of "furtherances" in the flow of our experience.

However, whenever experiences are discrepant with our prior expectancies, we are blocked and frustrated, unsure what to do. Our situation becomes indeterminate and more uncertain. We can no longer proceed under the guidance of prior impulses and habits. This is not just a frustration to be avoided, but rather the challenge of new evidence that engages focused attention and effortful reasoning. If we hope to manage our situation intelligently, we have to recalibrate our expectances so that they are more in line with our present experience and more adequately predict future experiences under changed conditions. Most of this reworking process goes on unconsciously. However, in some cases we are able to carry out this process at a conscious level, and when we make such conscious inquiries, we tend to think of these instances as knowledge in an eminent sense, even though the unconscious operations are also genuine modes of knowing. In short, human knowing begins with expectancies projected onto our experience of the world, to be readjusted, when necessary, in light of actual experiences.

9. Inquiry is a need-search-satisfaction process.

As Dewey astutely observed, there is no need for thought, or even conscious feeling, when we are moving along more or less effortlessly in life, in channels carved by our prior habits and expectancies. When our projected expectancies are satisfied by the affordances of our present experience, we feel the path opening before us. But when we encounter new conditions for which our prior habits have not prepared us, we experience the resulting indeterminacy of our situation as *need* for corrective action. When acting intelligently, this blockage, this frustration, this need is the spur for inquiry, which is the *search* for a way to resolve the indeterminacy by either readjusting our prior expectancies or generating new ones. When this process of inquiry is successful, we experience the *satisfaction* state of resolved tension and the release of energy into new activities. The need-search-satisfaction cycle emerges from our unconscious allostatic organic processes of adaptation, but it often then generates our conscious activities of problem solving and reflective inquiry. This is one important sense in which there is continuity between lower (i.e., sensory-motor, visceral) operations and higher (i.e., conceptualization and reasoning) forms of cognition. Our higher cognitive functions recruit structures and processes of our bodily sensory and motor operations to constitute our higher-level understanding and inferential reasoning.

10. Knowing is a personally transformative activity, and not just a representational account of some aspect of a mind-independent world.

The history of philosophy's attempts to characterize theoretical knowledge is mostly a narrative of efforts to explain how what is internal to mind mirrors external mind-independent states of affairs in the world. According to this tradition, to know is to represent, in some inner realm, the truth of how things are in the outer world. The radical alternative to this received view was first glimpsed in part by Kant (who argued that we do not just report on experience but actually co-constitute it), then articulated in pragmatist philosophy (which took knowing to be an activity that transforms experience), and later supported and extended by the cognitive science of the embodied mind (which explained how bodily schemas, feelings, emotions, motive controls, and values underlie our experience and our knowledge of it). This new view recognizes that knowing is primarily a mode of *action* that transforms our experience, rather than merely representing aspects of a fixed and finished world. The process of knowing is then a co-constitution of self and world because significant new learning requires a transformation of the self as we recalibrate our expectancies and revise certain habits. There are important representational dimensions of our knowing activity, but the crucial process is one of remaking experience—a mode of *acting* constructively in and on the world, guided by our motive control systems. The world as known through a process of inquiry becomes a different world from that which, a few moments earlier, was indeterminate and unstable. Because the self participates in this newly known world, it must grow accordingly. Knowing changes the meaning of things, and thereby changes both self and world.

11. *The construction of experience through knowing is an ongoing actualization of identity.*

This implication follows from the previous point. Knowledge—which is learning that enables us to function more intelligently in our world—is the implicit constitution of identity. We are bundles of interconnected habits—the expectancies we acquire through our ongoing experience over time. When we transform some of our expectations and habits, we transform our selfhood—who we are. The experience of objects is, at the same time, a constitution of the self that experiences those objects. Self and object are correlative dimensions of one continuous experiential process of organism-environment transactions. Since mind and world are correlative, interdependent realities, the activity of knowing is also a reworking of the self. The prior self is remade by restructuring, through the need-search-satisfaction process of inquiry, some of its expectations and habits of action. Learning changes the self, as it quite literally changes our brains. The stability-plasticity dilemma, whereby learning something significant means we must change, is a fundamental fact in the life of the mind.

12. *Elation and anxiety define two different, and complementary, modes of knowing.*

We saw that the process of knowing is mediated unconsciously by the adaptive controls of dual limbic circuits with specific constraints on neural activity in time and specific qualities of feeling. One circuit gives rise to elation (emphasizing the personal confidence that leads to impulsive action) and the other circuit gives rise to anxiety (vigilance for external errors and constraint that arises from lack of confidence in personal expectancies). When things go well, and there is no strong separation of self and world, we experience the mild *elation* of relative harmony with our surroundings and a corresponding flow in the process of experience. The world is our oyster. This pleasurable feeling of elation is phasic in character (i.e., a short-lived arousal that quickly habituates and fades). This is a form of knowing, in the sense that the organism can act relatively successfully in its world without encountering strain, blockage, or disharmony. The problem with this mode of action and its attendant pleasurable mood is that it has no critical dimension because it has no way to disengage partially from its ongoing experience in order to take stock of how things are going. Consequently, there is little genuine learning going on, and there is no basis for critical self-reflection. Lacking a component of critical assessment, this may not seem very much like knowledge, but it is nonetheless a form of tacit knowing insofar as it makes use of prior experience to negotiate present action.

The other mode of knowing arises as a need-search-satisfaction process in response to the indeterminacy of situations in which prior projected expectancies fail to mesh with what we actually experience. The temporary blockage that we experience is an opportunity for sufficient distance between self and situation. We hesitate, and pull back, so that critical reflection, with its reconstructive possibilities, becomes possible. The initial encounter with discrepant experience is an indeterminacy that generates *anxiety* and its attendant feelings (e.g., fear). This anxious state is accompanied by a more narrowly focused heightened attention to an object (i.e., sensitization) that extends over a longer period of arousal (i.e., it is tonic in character). This intensified focus stimulates a search for a way to resolve the discrepancy between prediction and experience. This second mode of knowing, with its anxious mood, is a more critically informed and potentially reflective way of solving problems. We gain critical perspective as we engage in exploratory inquiry, but we suffer from anxiety and uncertainty along the way, which can sometimes inhibit appropriate action.

There are inherent structural influences in the process of conceptualization that result from the adaptive constraints of elation and anxiety on the process of memory consolidation and concept formation. The sensitization bias of the ventral limbic circuits creates a form of motivational control that restricts the scope of neural activity (by focusing on selected objects) but thereby extends its constancy in time. This explains the focus of attention and analytic (object) perception within an anxious state that can extend temporally over the entire course of the inquiry. Everyone has, from time to time, experienced this anxious sense that something is wrong and you don't know what to do or how to resolve your problematic situation. If you don't just get overwhelmed, the motive control of anxiety causes you to worry, to focus on the threat, and to obsess with finding a solution until you are more or less effectively coping with the problem.

This anxious, highly critical, and analytic mode of knowing contrasts sharply with the habituation bias of the dorsal limbic circuits that results in an expansive scope of neural representation, but at the expense of constancy in time. This explains the holistic scope of spatial (contextual) memory that is engaged in an elated mood state. It is the unreflective flow of experience when there is no awareness of indeterminacy, frustration, or doubt.

13. *The elated mode of experience is uncritical, self-aggrandizing, and extraverted, whereas the anxious mode is critical, self-critical, and introverted. Exaggeration of either mode can lead to a personality disorder, involving systematic distortions of cognition.*

The two modes of adaptive limbic control shape not only the immediate control of cognition but also the ongoing organization of the personality. When either is exaggerated and unbalanced, the result is a personality disorder—a deficit of self-regulation in the personality, tending toward impulsivity and extraversion, on the one hand, and constraint and introversion on the other. At one extreme, we see the grandiosity of mania, the global, extroverted impressionistic type of cognition, or, similarly, the impulsive life choices of the histrionic personality. At the other extreme, we see the excessive constraint and error checking (i.e., introverted analytic excess) of the person with obsessive-compulsive disorder or, similarly, the social alienation and egocentric suspicion characteristic of the paranoid personality.

Not only is our personal conceptual structure formed by our feelings, but the adaptive constraints of anxiety and elation remain concretizing influences when they operate unconsciously, particularly when they are so strong as to overwhelm and unbalance one's personality. Both modes are egocentric, and only through balancing them do we emerge from egocentrism to appreciate the needs and perspective of other people in our lives. Through this appreciation we learn to internalize the perspectives of others so that knowing engages the evidence of the world in a way that allows knowledge to become intersubjective and objective rather than mere wishful projection.

14. *The capacity for increased self-awareness, gained primarily through understanding others' perspectives, gives rise to abstract thinking.*

Abstraction is an ability to pull back from the immediacy of the self's fusion with the concrete objects of experience in order to see those objects in relation to other objects and events (past, present, and future). This reflective process requires self-awareness, so that knowing and identity are not so implicitly fused, and knowing becomes more objective. Self-awareness, in turn, requires intersubjective awareness that other people have their own perspectives, just as you have yours. This ability to take up the perspective of another makes it possible to now regard one's own preferred organization of experience as itself just one possible perspective. We can then step back and hold that perspective up for scrutiny and evaluation. We become able to hold our present perspective or experience at arm's length, interrogating its sources and its consequences for experience. This ability to relate a present object to objects distant in space and/or time is abstraction. We abstract some characteristic or relation that we then discover to be operative in other situations, which constitutes an experience of new meaning. The self-awareness that accompanies abstract thought is the basis for hypothetical (i.e., as-if) reasoning, through separating the self from objects within its immediate experience. Reflective consciousness—achieved primarily through critical self-awareness—allows metacognition, a higher-order form of knowing. In this process, because of intersubjective experiences, we are no longer egocentrically (or ethnocentrically) fused with the process of knowing. Self-actualization allows knowing to be more tentative, more objective (i.e., intersubjectively validated), and less egocentric.

Only by standing back a bit, partially disengaging from the immediacy of present experience, can we see ourselves as entertaining alternative conceptualizations and interpretations, but without yet directly committing to one over another. As a result, we grow more allostatic (anticipating unexpected experiences) and less homeostatic (constrained by preset equilibrium states). With this distance from the process of cognition provided by self-awareness, we are not so fused with the process of knowing and can entertain knowledge without having to define the self entirely anew with every instance of reflective thought.

15. Conceptualization is abstraction; concepts are expectancies of affordances.

The mistaken notion that knowledge, in its eminent sense, is expressed only through concepts and propositions has fatefully influenced Western epistemology. Concepts are not inner representations mirroring external states of affairs, nor are they entities of any sort, concrete or abstract. If concepts can be said to be *in* something, that something would *not* be a mysterious realm of mind, but rather a brain, in a living body, interacting with a changing environment. There is nowhere for concepts to be except as patterns of neural connectivity that are activated by ongoing experiences.

Concepts arise from our ability to select, mostly unconsciously, characteristics and relations from the much-at-once-ness of everyday experience (James 1911) and then to see how those properties and relations have meaning and are operative in other objects or events. Knowing of this sort abstracts from the immediacy of a present concrete experience in order to see connections to other experiences, thereby expanding and enriching the meaning of an object or event. The process of concept formation is abstraction made possible by increasing self-awareness that frees us somewhat from our egocentric predicament. This pulling or standing back (self-awareness) allows us to see ourselves as having a perspective on what we selectively attend to. It allows us to see relations and connections that were not evident when we were immersed in the immediacy of our concrete holistic experience.

The process of abstraction is realized in the structure of the brain via relations among the primary sensory, secondary (unimodal) association, heteromodal association, and limbic regions. Each "higher" level—such as in moving from a unimodal association level to the activation of patterns of connectivity in the heteromodal association area—represents a higher level of abstraction. In this sense, all conceptualization is a form of abstraction, which means that the notion of concrete versus abstract concepts should be interpreted as a continuum rather than a dichotomy.

Concept formation is a process that combines differentiation (recognizing differences among members of a category) with integration (seeing those different individual objects as closely related and therefore unified by at least some shared features). We recognize differences among instantiations (e.g., different types of beetles) while at the same time they are integrated into a unified conceptual structure (e.g., the concept *beetle*). Having a concept is having a relatively unified set of expectancies (realized as patterns of neural connectivity) about future experiences in relation to some kind of thing or state of affairs. In short, a concept is a probability distribution of expectancies for future affordances (in relation to some object or event) based on affordances encountered in previous experiences. Fundamentally, a concept is a generalization of experience.

Emerging from its mammalian roots, the human brain has developed the basic capacities for spatial and object concepts in fundamental ways, including the specialization of right and left hemispheres to elaborate conceptual forms within major domains of mind. The dorsal and ventral modes of motive control have been extended to the specialized cerebral hemispheres, such that the management of certainty and uncertainty provide uniquely human ways of regulating creative generation of intuitions, on the one hand, and critical analysis and reasoning, on the other.

Through the left hemisphere's analytic and routinized capacities for language, paralleled by the right hemisphere's holistic representation of emotional and spatial context, human intelligence has been able to integrate the information complexity of generations of culture through language and other forms of symbolic communication (such as music, painting, architecture, gesture, and ritual). The analysis of the metaphors of abstract thought has provided evidence of how the mind is embodied in feelings, perceptions, and actions while still achieving abstract concepts that allow efficient and sophisticated cultural communication.

16. All knowledge is contextual.

There is no disembodied knowledge. We cannot jump out of our skins to achieve a perspective-free, all-knowing grasp of the nature of things. The need to know arises from the indeterminacy of an organism's encounter with some parts of its environment, and this need is always shaped by our visceral homeostatic and allostatic processes and modulated by our embodied motivational systems. Our embeddedness in the world is not a problem to be overcome, but rather the necessary condition of knowing that is relevant to our existence.

The experiential context, which Dewey called the *situation*, circumscribes the phenomena we seek to understand, and it establishes which factors are relevant to achieving understanding adequate for our present purposes. In other words, without a context, we don't know what problem we have encountered, what it is that we are trying to make sense of, or how to recognize when we have gained knowledge through our inquiries. So, there can be no knowledge without a context—a context shaped by our visceral needs and values as we engage our surroundings to preserve our life and to enhance the quality of our existence.

17. *There is no absolute, final knowledge.*

Because all life activity goes on through intimate interaction with a nature that is both relatively stable but also subject to changes resulting from newly emergent conditions, learning is never finished, so long as we draw breath. *Learning is cognition extend over the life span of the organism*—an ongoing reconstructing of self and world. The time is past to hold out for certain, final, and absolute knowledge. We never had such transcendent knowledge, and the truth is that we never really needed any such thing in order to survive and flourish. What we actually need is a form of knowing situated in our bodily transactions with our world, in such a way that we can be relatively "in touch with" and "at home in" our physical, interpersonal, and cultural surroundings. We need a critical capacity to understand the meaning of our situation and to enact intelligently guided change. The challenge for the sciences and humanities should not be to find some illusory unchanging truth, but rather to engage conscientiously in critical reflective inquiry geared toward resolution of problems that we encounter in our lives. "Being at home" in the world does not mean merely settling into one's preferred habits and drifting unreflectively, the way most of us operate most of the time. The "world" that we should strive to be at home in is not our parochial situation, but rather the widest range of relevant conditions within which we find ourselves and which we can in some way transform through our inquiries. The goal of a humanistic liberal education rooted in the sciences, humanities, and arts is to develop the critical and communicative skills necessary to achieve the broadest and most critically astute perspective geared toward making things better.

18. *Knowing can be both objectively validated and personally meaningful.*

We come full circle back to the question about the possibility of a meaningful mind science. We argued that the psychology and first-generation cognitive science of the twentieth century were pursuing notions of pristine objectivity that are mistaken, unattainable, and unnecessary. Fortunately, we are not condemned to replicate the two cultures of the academy that have plagued us for decades. We have developed methods and technologies of inquiry—the methods and techniques of the sciences, humanities, and arts—that can be employed by others and used to validate, or invalidate, research results objectively and intersubjectively. At the same time, we have an expanding and deepening awareness of the subjective dimensions of our cognition, which is giving us a better appreciation of how knowledge can be meaningful and relevant to our lives. What we are discovering about the workings of mind can be meaningful to us insofar as it helps us know who we are and how we structure our understanding of everything. We can learn to appreciate what a biologically based second-generation cognitive science reveals about the vast unconscious processes of mind and knowing, and we can try to guide our knowing practices more reflectively in light of these profound insights.

12.3   Who Are We? Knowing in the Embodied Mind

The challenge of understanding ourselves through a natural philosophy requires that we appreciate both objective and subjective perspectives, and thereby reject the division of the two cultures of humanism and science. Neither side of this division can be accepted as adequate for educating the embodied mind in the information age. The Delphic imperative to "Know Thyself" takes on special meaning in the context of the Information Age. Otherwise, the decisions on developing and implementing information technology will be made by those whose intelligence is narrowly technical, and who may have their own values, goals, and agendas for our lives.

In contrast, the emerging opportunity for a natural philosophy suggests new perspectives on what makes us who we are, how we understand things, where our values come from, and how we are able to gain reliable knowledge of ourselves and our world. We can see how our ability to learn about the world—which is an ongoing process of reconstruction and growth—is rooted in the nature of our bodies, brains, and the environments we inhabit. Instead of assuming a disembodied mind or transcendent ego, we can explain the workings of everything we count as mind in terms of our body-based trans-actions with our natural and social worlds, and therefore without any need to posit supernatural or transcendent entities, forces, processes, and values.

Together, science and philosophy are offering a new view of the natu-ral basis of the mind. This is not the ideal plane of pristine forms or ideas that Plato envisioned outside of his cave. Classical idealism was an illusion based on reifying abstract concepts, not unlike the modern philosopher's fallacy of taking our linguistic distinctions as capturing the essential nature of things. Our refiguring of Plato's famous analogy regards the world outside the cave as the natural world without illusions, where the mind is discovered to operate through natural events and processes evolved over millennia, developing over our lives, and firmly rooted in our embodiment and culture.

So, who are we? As Dewey (1925, ch. 10) reminded us, we are not "little gods" who bring universal reason, pure ego, pristine objectivity, and absolute knowledge to bear on our world from some transcendent perspective. Instead, we are inescapably *in* and *of* the world—complex social animals who cannot shed our skins, but whose animality is the very source of our highest cognitive achievements. Animals, yes, but remarkable animals who recruit from our mammalian heritage capacities for perception, motive control, emotions, and action, and use them knowingly for negotiating our way through life. Our cognition is built from the bottom up, growing from our biological capacities and values, but that does not leave us in the immanence of bodily engagement and animal life. Through the ever-increasing complexity of our bodies, brains, and environments, we achieve new levels of functional organization that make possible unique human understanding and knowing, even as they are rooted in our basic organism-environment interactions. The evolutionary gift of self-awareness is the key to our higher forms of knowing. Those activities of knowing can be both intersubjectively valid and also personally meaningful, precisely because they are shaped by our deepest motivations and values.

Therefore, when you really come to appreciate the inescapable and profound role of our embodiment, you are reminded that you are, after all, an animal struggling to survive, flourish, and find meaning in a material, interpersonal, and cultural world. You are the most recent instantiation of a remarkable process of biological evolution. This is humbling but also comforting. It is humbling insofar as it denies any radical distinction between humans and other animals, emphasizing instead what we share with many of our animal ancestors. It is comforting insofar as it shows us that we are mostly in touch with our surroundings, as we are embedded in nature, and have developed cultural strategies for growing and learning that can lead to genuinely useful knowledge. It is inspiring insofar that it also recognizes some distinctive human capacities for knowing, intelligent, creative engagement with our world. We are, quite obviously, imperfect animals, messing things up in all sorts of ways and disappointingly falling short of our imagined ideals. But perfection is an empty, unattainable, and unnecessary illusion. What we need is viability—only to be "good enough"—and we do, indeed, have adequate capacities and resources for accomplishing that on many occasions. When we fall we can pick ourselves up, dust ourselves off, and try, try again. Trying again is not a fool's errand because our modest faith in intelligence has frequently been validated by some of our most impressive achievements in knowing our world.

To the extent that we are special, it is not because we radically transcend our animality, but rather because we are the beneficiaries of mammalian capacities that, under the influence of increased complexity of cultural transmission, give rise to emergent levels of functional organization, and cumulative forms of cultural intelligence, that are then indeed unique to the human species and that help us live more intelligent and worthwhile lives. There is then philosophical significance in understanding our continuity—in the immediate mechanisms of experience—with the hundreds of millions of years of vertebrate evolution.

References

Aggleton, J. P., & Mishkin, M. (1986). The amygdala: Sensory gateway to the emotions. In R. Plutchik & H. Kellerman (Eds.), *Emotion: Theory, research and experience* (Vol. 3, pp. 281–299). New York: Academic Press.

Aksnes, H. (2002). *Perspectives of musical meaning: A study based on selected works by Geirr Tveitt* (Doctoral dissertation). Faculty of Arts, University of Oslo, Norway.

Amaral, D. G., Price, J. L., Pitkänen, A., & Carmichael, S. T. (1992). Anatomical organization of the primate amygdaloid complex. In J. P. Aggleton (Ed.), *The amygdala* (pp. 1–66). New York: Wiley-Liss.

American Psychiatric Association. (2013). *Diagnostic and statistical manual of mental disorders* (5th ed.). Washington, DC: American Psychiatric Publishing.

Andreasen, N. C. (2008). The relationship between creativity and mood disorders. *Dialogues in Clinical Neuroscience, 10*(2), 251–255.

Andreasen, N. J. C., & Canter, A. (1974). The creative writer: Psychiatric symptoms and family history. *Comprehensive Psychiatry, 15*, 123–131.

Andreasen, N. J. C., & Powers, P. S. (1975). Creativity and psychosis: An examination of conceptual style. *Archives of General Psychiatry, 32*, 70–73.

Aristotle. (2009a). *Nicomachean ethics*. New York: Modern Library.

Aristotle. (2009b). *Posterior analytics*. New York: Modern Library.

Astafiev, S. V., Shulman, G. L., Stanley, C. M., Snyder, A. Z., Van Essen, D. C., & Corbetta, M. (2003). Functional organization of human intraparietal and frontal cortex for attending, looking, and pointing. *Journal of Neuroscience, 23*(11), 4689–4699. doi: 10.1523/JNEUROSCI.23-11-04689.2003.

Aston-Jones, G., & Cohen, J. D. (2005). Adaptive gain and the role of the locus coeruleus-norepinephrine system in optimal performance. *Journal of Comparative Neurology, 493*(1), 99–110. Retrieved from https://pubmed.ncbi.nlm.nih.gov/16254995.

Aston-Jones, G., Ennis, M., Pieribone, V. A., Nickell, W. T., & Shipley, M. T. (1986). The brain nucleus locus coeruleus: Restricted afferent control of a broad efferent network. *Science, 234*, 734–737.

Baars, B. J. (1986). *A cognitive theory of consciousness.* New York: Cambridge University Press.

Bailey, D. (1997). *A computational model of embodiment in the acquisition of action verbs* (Doctoral dissertation). Computer Science Division, EECS Department, University of California, Berkeley.

Barbas, H. (1995). Anatomic basis of cognitive-emotional interactions in the primate prefrontal cortex. *Neuroscience & Biobehavioral Reviews, 19*(3), 499–510. Retrieved from https://pubmed.ncbi.nlm.nih.gov/7566750.

Barrett, L. F. (2017). The theory of constructed emotion: An active inference account of interoception and categorization. *Social Cognitive and Affective Neuroscience, 12*(1), 1–23. doi: 10.1093/scan/nsw154.

Barsalou, L. W. (1999). Perceptual symbol systems. *Behavioral and Brain Sciences, 22*(4), 577–660.

Bechtel, W. (2008). *Mental mechanisms: Philosophical perspectives on cognitive neuroscience.* New York: Psychology Press.

Bergen, B. K. (2012). *Louder than words: The new science of how the mind makes meaning.* New York: Basic Books.

Bhatt, R. (Ed.). (2013). *Rethinking aesthetics: The role of body in design.* New York: Routledge.

Black, M. (1954–1955). Metaphor. *Proceedings of the Aristotelian Society,* n.s., *55,* 273–294.

Blakemore, S. J., & Frith, U. (2004). How does the brain deal with the social world? *Neuroreport, 15*(1), 119–128. Retrieved from https://pubmed.ncbi.nlm.nih.gov/15106843.

Blumer, D., & Benson, D. F. (1975). Personality changes with frontal and temporal lobe lesions. In D. F. Benson & D. Blumer (Eds.), *Psychiatric aspects of neurologic disease* (pp. 151–170.). New York: Grune and Stratton.

Boroditsky, L. (2011). How languages construct time. In S. Dehaene & E. Brannon (Eds.), *Space, time, and number in the brain: Searching for the foundations of mathematical thought* (pp. 333–341). London: Elsevier.

Brown, J. W. (1994). Morphogenesis and the mental process. *Development and Psychopathology, 6,* 551–563.

Brown, N. O. (1959). *Life against death: The psychoanalytical meaning of history.* New York: Vintage Books.

Brugman, C. (1983). The use of body-part terms as locatives in Chalcatongo Mixtec. In A. Schlichter et al. (Eds.), *Reports from the survey of California and other Indian languages*, 235–290. Berkeley: University of California, Berkeley.

Cartwright, R. D. (2004). The role of sleep in changing our minds: A psychologist's discussion of papers on memory reactivation and consolidation in sleep. *Learning & Memory, 11*(6), 660–663. doi: 10.1101/lm.75104.

Cartwright, R. D., Agargun, M. Y., Kirkby, J., & Friedman, J. K. (2006). Relation of dreams to waking concerns. *Psychiatry Research, 141*(3), 261–270. doi: 10.1016/j.psych res.2005.05.013.

Carver, C. S., & Scheier, M. (1990). Origins and functions of positive and negative affect: A control-process view. *Psychological Review, 97,* 19–35.

Changeux, J., & Dehaene, S. (1989). Neuronal models of cognitive functions. *Cognition, 33,* 63–109.

Churchland, P. S. (1986). *Neurophilosophy: Toward a unified science of the mind-brain.* Cambridge, MA: MIT Press.

Churchland, P. S. (2002). *Brain-wise: Studies in neurophilosophy.* Cambridge, MA: MIT Press.

Clark, A. (2015). *Surfing uncertainty: Prediction, action, and the embodied mind.* New York: Oxford University Press.

Corbetta, M., Patel, G., & Shulman, G. L. (2008). The reorienting system of the human brain: From environment to theory of mind. *Neuron, 58*(3), 306–324. doi: 10 .1016/j.neuron.2008.04.017.

Damasio, A. (1994). *Descartes' error: Emotion, reason, and the human brain.* New York: G. P. Putnam's Sons.

Damasio, A. (1998). Emotion in the perspective of an integrated nervous system. *Brain Research Reviews, 26*(2–3), 83–86. Retrieved from https://pubmed.ncbi.nlm.nih .gov/9651488.

Damasio, A. (1999). *The feeling of what happens: Body and emotion in the making of consciousness.* New York: Mariner Books.

Damasio, A. (2003). *Looking for Spinoza: Joy, sorrow, and the feeling brain.* Orlando, FL: Harcourt.

Damasio, A. (2010). *Self comes to mind: Constructing the conscious brain.* New York: Vintage Books.

Damasio, A. (2018). *The strange order of things: Life, feeling, and the making of cultures.* New York: Pantheon Books.

Damasio, A. R., Tranel, D., & Damasio, H. (1990). Face agnosia and the neural substrates of memory. *Annual Review of Neuroscience, 13*, 89–109.

Dancygier, B., & Sweetser, E. (2014). *Figurative language*. New York: Cambridge University Press.

Dehaene, S. (2014). *Consciousness and the brain: Deciphering how the brain codes our thoughts*. New York: Viking.

Dehaene, S., & Changeux, J. P. (2011). Experimental and theoretical approaches to conscious processing. *Neuron, 70*(2), 200–227. doi: 10.1016/j.neuron.2011.03.018.

Dehaene, S., Changeux, J. P., Naccache, L., Sackur, J., & Sergent, C. (2006). Conscious, preconscious, and subliminal processing: A testable taxonomy. *Trends in Cognitive Sciences, 10*(5), 204–211. doi: 10.1016/j.tics.2006.03.007.

Dehaene, S., Lau, H., & Kouider, S. (2017). What is consciousness, and could machines have it? *Science, 358*(6362), 486–492.

Derryberry, D., & Tucker, D. M. (1991). The adaptive base of the neural hierarchy: Elementary motivational controls on network function. In A. Dienstbier (Ed.), *Nebraska symposium on motivation* (pp. 289–342). Lincoln: University of Nebraska Press.

Derryberry, D., & Tucker, D. M. (2006). Motivation, self-regulation, and self-organization. In D. J. Cohen & D. Cicchetti (Eds.), *Handbook of developmental psychopathology* (Vol. 2, *Developmental neuroscience*, pp. 502–532). New York: Wiley.

Descartes, R. (1628/1970). *Rules for the direction of the mind*. In *The philosophical works of Descartes* (Vol. 1, pp. 1–79) (E. Haldane and G. R. T. Ross, Trans.). Cambridge: Cambridge University Press.

Descartes, R. (1637/1970). *Discourse on method*. In *The philosophical works of Descartes* (Vol. 1, 79–130) (E. Haldane and G. R. T. Ross, Trans.). Cambridge: Cambridge University Press.

Dewey, J. (1896). The reflex arc concept in psychology. *Psychological Review, 3*(4), 357–370.

Dewey, J. (1922/1988). *Human nature and conduct*. In *The middle works, 1899–1924* (Vol. 14) (J. Boydston, Ed.). Carbondale: Southern Illinois University Press.

Dewey, J. (1925/1981). *Experience and nature*. In *The later works, 1925–1953* (Vol. 1) (J. Boydston, Ed.). Carbondale: Southern Illinois University Press.

Dewey, J. (1929/1984). *The quest for certainty*. In *The later works, 1925–1953* (Vol. 4) (J. Boydston, Ed.). Carbondale: Southern Illinois University Press.

Dewey, J. (1930/1981). *Qualitative thought*. In *The later works, 1925–1953* (Vol. 5) (J. Boydston, Ed.). Carbondale: Southern Illinois University Press.

Dewey, J. (1934/1987). *Art as experience*. In *The later works, 1925–1953* (Vol. 10) (J. Boydston, Ed.). Carbondale: Southern Illinois University Press.

Dodge, E., & Lakoff, G. (2005). Image schemas: From linguistic analysis to neural grounding. In B. Hampe (Ed.), *From perception to meaning: Image schemas in cognitive linguistics* (pp. 57–91). Berlin: Mouton de Gruyter.

Edelman, G. (1987). *Neural Darwinism: The theory of neuronal group selection*. New York: Basic Books.

Edelman, G. (1989). *The remembered present: A biological theory of consciousness*. New York: Basic Books.

Edelman, G., & Tononi, G. (2000). *A universe of consciousness: How matter becomes imagination*. New York: Basic Books.

Eidelberg, D., & Galaburda, A. M. (1984). Inferior parietal lobule: Divergent architectonic asymmetries in the human brain. *Archives of Neurology, 41*(8), 843–852. Retrieved from https://pubmed.ncbi.nlm.nih.gov/6466160.

Fauconnier, G., & Turner, M. (2002). *The way we think: Conceptual blending and the mind's hidden complexities*. New York: Basic Books.

Feldman, J. (2006). *From molecule to metaphor: A neural theory of language*. Cambridge, MA: MIT Press.

Fernandez-Duque, D., and Johnson, M. (1999). Attention metaphors: How metaphors guide the cognitive psychology of attention. *Cognitive Science, 23*(19), 83–116.

Fernandez-Duque, D., and Johnson, M. (2002) Cause and effect theories of attention: The role of conceptual metaphors. *Review of General Psychology, 6*(2), 153–165.

Fichte, J. G. (1796/2000). *Foundations of natural right* (Michael Baur, Trans.). Cambridge: Cambridge University Press.

Floridi, L. (2014). *The 4th revolution: How the infosphere is reshaping human reality*. New York: Oxford University Press.

Forceville, C. (2009). Non-verbal and multimodal metaphor in a cognitivist framework: Agendas for research. In C. J. Forceville & E. Urios-Aparisi (Eds.), *Multimodal metaphor* (pp. 19–35). Berlin: Walter de Gruyter.

Forceville, C. J., & Urios-Aparisi, E. (Eds.). (2009). *Multimodal metaphor*. Berlin: Walter de Gruyter.

Frege, G. (1966). On sense and reference. In P. Geach and M. Black (Eds.), *Translations from the philosophical writings of Gottlob Frege*. Oxford: Basil Blackwell.

Freud, S. (1895). Project for a scientific psychology. In J. Strachey (Ed.), *The standard edition of the complete psychological works of Sigmund Freud* (Vol. 1, pp. 295–344). London: Hogarth Press, 1966.

Freud, S. (1940). *An outline of psychoanalysis*. In J. Strachey (Ed.), *The standard edition of the complete psychological works of Sigmund Freud* (Vol. 23). London: Hogarth Press, 1964.

Freud, S. (1953). *The interpretation of dreams*. London: Hogarth Press (first German edition 1900).

Friston, K. (2018a). Am I self-conscious? (Or does self-organization entail self-consciousness?). *Frontiers in Psychology, 9*, 579. doi: 10.3389/fpsyg.2018.00579.

Friston, K. (2018b). Does predictive coding have a future? *Nature Neuroscience, 21*(8), 1019–1021. doi: 10.1038/s41593-018-0200-7.

Galaburda, A. M., & Pandya, D. N. (1983). The intrinsic architectonic and connectional organization of the superior temporal region of the rhesus monkey. *Journal of Comparative Neurology, 221*(2), 169–184. Retrieved from https://pubmed.ncbi.nlm.nih.gov/6655080.

Gallagher, S. (2005). *How the body shapes the mind*. New York: Oxford University Press.

Gallese, V., & Lakoff, G. (2005). The brain's concepts: The role of the sensory-motor system in conceptual knowledge. *Cognitive Neuropsychology, 22*(3–4), 455–479.

García-Cabezas, M. Á., Zikopoulos, B., & Barbas, H. (2019). The Structural Model: A theory linking connections, plasticity, pathology, development and evolution of the cerebral cortex. *Brain Structure and Function, 224*(3), 985–1008. doi: 10.1007/s00429-019-01841-9.

Gibbs, R. W., Jr. (1994). *The poetics of mind: Figurative thought, language, and understanding*. Cambridge: Cambridge University Press.

Gibbs, R. W., Jr. (2005). *Embodiment and cognitive science*. New York: Cambridge University Press.

Gibbs, R. W., Jr. (Ed.). (2008). *The Cambridge handbook of metaphor and thought*. Cambridge: Cambridge University Press.

Gibson, J. J. (1950). *The perception of the visual world*. Boston: Houghton Mifflin.

Gibson, J. J. (1979). *The ecological approach to visual perception*. Boston: Houghton Mifflin.

Givon, T. (2002). *Bio-Linguistics*. Amsterdam: John Benjamins.

Goldberg, A. (1995). *Constructions: A construction grammar approach to argument structure*. Chicago: University of Chicago Press.

Goldberg, G. (1985). Supplementary motor area structure and function: Review and hypotheses. *Behavioral and Brain Sciences, 8*, 567–616.

Goldberg, G., & Bloom, K. K. (1990). The alien hand sign: Localization, lateralization and recovery. *American Journal of Physical Medicine & Rehabilitation, 69*(5), 228–238. Retrieved from https://pubmed.ncbi.nlm.nih.gov/2222983.

Goldstein, K. (1939). *The organism: A holistic approach to biology derived from pathological data in man*. New York: American Book Company.

Goldstein, K. (1952). The effect of brain damage on the personality. *Psychiatry, 15*, 245–260.

Goodale, M. A., & Milner, A. D. (1992). Separate visual pathways for perception and action. *Trends in Neuroscience, 15*(1), 20–25. doi: 10.1016/0166-22236(92)90344-8.

Gould, S. J. (1977). *Ontogeny and phylogeny*. Cambridge, MA: Harvard University Press.

Grady, J. (1997). *Foundations of meaning: Primary metaphors and primary scenes*. Ph.D. dissertation, University of California, Berkeley.

Grossberg, S. (1980). How does a brain build a cognitive code? *Psychological Review, 87*(1), 1–51.

Grube, G. M., & Reeve, C. D. C. (1974). *Plato's republic*. Indianapolis, IN: Hackett Publishing Company.

Harvey, O. J., Hunt, D. E., & Schroder, H. M. (1961). *Conceptual systems and personality organization*. New York: Wiley.

Hebb, D. O. (1949). *The organization of behavior*. New York: Wiley.

Heims, S. J. (1991). *The cybernetics group*. Cambridge, MA: MIT Press.

Hempel, C. (1965). *Aspects of scientific explanation*. New York: Free Press.

Hohwy, J. (2014). *The predictive mind*. New York: Oxford University Press.

Hume, D. (1739). *A treatise on human nature*. Oxford: Clarendon Press.

Ichikawa, J. J., & Steup, M. (2018). The analysis of knowledge. In E. N. Zalta (Ed.), *The Stanford encyclopedia of philosophy* (Summer ed.). https://plato.stanford.edu/archives/sum2018/entries/knowledge-analysis/.

Jackson, P. L., Meltzoff, A. N., & Decety, J. (2005). How do we perceive the pain of others? A window into the neural processes involved in empathy. *Neuroimage, 24*(3), 771–779. Retrieved from https://pubmed.ncbi.nlm.nih.gov/15652312.

James, W. (1890/1950). *The principles of psychology*. 2 vols. New York: Dover.

James, W. (1909/2002). *The meaning of truth*. New York: Dover.

James, W. (1911). Percept and concept. In *Some problems of philosophy: A beginning of an introduction to philosophy* (pp. 47–74). New York: Longmans, Green and Co.

Johnson, M. (1987). *The body in the mind: The bodily basis of meaning, imagination, and reason*. Chicago: University of Chicago Press.

Johnson, M. (2007). *The meaning of the body: Aesthetics of human understanding*. Chicago: University of Chicago Press.

Johnson, M. (2017). *Embodied mind, meaning, and reason: How our bodies give rise to understanding*. Chicago: University of Chicago Press.

Johnson, M. (2018). *The aesthetics of meaning and thought: The bodily roots of philosophy, science, morality, and art*. Chicago: University of Chicago Press.

Kant, I. (1781/1968). *Critique of pure reason* (N. K. Smith, Trans.). New York: St. Martin's Press.

Kant, I. (1788/2002). *Critique of practical reason*. Indianapolis, IN: Hackett Publishing Company.

Kant, I. (1790/1987). *Critique of judgment*. Indianapolis, IN: Hackett Publishing Company.

Kohut, H. (1978). *The search for the self*. New York: International Universities Press.

Kovecses, Z. (2010). *Metaphor: A practical introduction*. New York: Oxford University Press.

Kovecses, Z. (2020). *Extended conceptual metaphor theory*. Cambridge: Cambridge University Press.

Kragh, U., & Smith, G. J. W. (Eds.). (1970). *Percept-genetic analysis*. Lund, Sweden: Gleerup.

Kuhn, T. (1962). *The structure of scientific revolutions*. Chicago: University of Chicago Press.

Lakoff, G. (1987). *Women, fire, and dangerous things: What categories reveal about the mind*. Chicago: University of Chicago Press.

Lakoff, G. (2008). The neural theory of metaphor. In R. W. Gibbs Jr. (Ed.), *The Cambridge handbook of metaphor and thought* (pp. 17–38). Cambridge: Cambridge University Press.

Lakoff, G., & Johnson, M. (1980). *Metaphors we live by*. Chicago: University of Chicago Press.

Lakoff, G., & Johnson, M. (1999). *Philosophy in the flesh: The embodied mind and its challenge to Western thought*. New York: Basic Books.

Lakoff, G., & Narayanan, S. (in press). *The neural mind*.

Lakoff, G., & Núñez, R. E. (2000). *Where mathematics comes from: How the embodied mind brings mathematics into being*. New York: Basic Books.

Lakoff, G. & Turner, M. (1989). *More than cool reason: A field-guide to poetic metaphor*. Chicago: University of Chicago Press.

Langacker, R. W. (1986). An introduction to cognitive grammar. *Cognitive Science, 10*(1), 1–40.

Langacker, R. W. (1987–1991). *Foundations of cognitive grammar* (2 vols.). Stanford, CA: Stanford University Press.

Langacker, R. W. (2002). *Concept, image, and symbol: The cognitive basis of grammar* (2nd ed.). Berlin: Mouton de Gruyter.

Liu, D., Diorio, J., Tannenbaum, B., Caldji, C., Francis, D., Freedman, A., . . . Meaney, M. J. (1997). Maternal care, hippocampal glucocorticoid receptors, and hypothalamic-pituitary-adrenal responses to stress. *Science, 277,* 1659–1662.

Luu, P., Jiang, Z., Poulsen, C., Mattson, C., Smith, A., & Tucker, D. M. (2011). Learning and the development of contexts for action. *Frontiers in Human Neuroscience, 5,* 159. doi: 10.3389/fnhum.2011.00159.

Luu, P., & Tucker, D. M. (2001). Regulating action: Alternating activation of midline frontal and motor cortical networks. *Clinical Neurophysiology, 112*(7), 1295–1306. Retrieved from https://pubmed.ncbi.nlm.nih.gov/11516742.

Luu, P., & Tucker, D. M. (2003a). Self-regulation and the executive functions: Electrophysiological clues. In A. Zani & A. M. Preverbio (Eds.), *The cognitive electrophysiology of mind and brain* (pp. 199–223). San Diego, CA: Academic Press.

Luu, P., & Tucker, D. M. (2003b). Self-regulation by the medial frontal cortex: Limbic representation of motive set-points. In M. Beauregard (Ed.), *Consciousness, emotional self-regulation and the brain* (pp. 123–161). Amsterdam: John Benjamins.

Luu, P., Tucker, D. M., & Derryberry, D. (1998). Anxiety and the motivational basis of working memory. *Cognitive Therapy and Research, 22,* 577–594.

Luu, P., Tucker, D. M., Derryberry, D., Reed, M., & Poulsen, C. (2003). Electrophysiological responses to errors and feedback in the process of action regulation. *Psychological Science, 14*(1), 47–53.

Madzia, R., & Jung, M. (2016). *Pragmatism and embodied cognitive science: From bodily intersubjectivity to symbolic articulation.* Berlin: Walter de Gruyter.

Marcuse, H. (1941). *Reason and revolution: Hegel and the rise of social theory.* Atlantic Highlands, NJ: Humanities Press.

Marin-Padilla, M. (1998). Cajal-Retzius cells and the development of the neocortex. *Trends in Neurosciences, 21*(2), 64–71.

Maslow, A. (1968). *Toward a psychology of being.* New York: Wiley.

Masson, J. M. (2003). *The assault on truth: Freud's suppression of seduction theory.* New York: Random House.

McClelland, J. L., McNaughton, B. L., & O'Reilly, R. C. (1995). Why there are complementary learning systems in the hippocampus and neocortex: Insights from the successes and failures of connectionist models of learning and memory. *Psychological Review, 102*, 419–457.

McCulloch, W. S., & Pitts, W. (1943). A logical calculus of the ideas immanent in nervous activity. *Bulletin of Mathematical Biophysics, 52*(1–2), 99–115. Retrieved from https://pubmed.ncbi.nlm.nih.gov/2185863.

McKenna, E. (2018). *Livestock: Food, fiber, and friends.* Athens: University of Georgia Press.

McKeon, R. (1941). *The basic works of Aristotle.* New York: Random House.

McNeill, D. (1992). *Hand and mind: What gestures reveal about thought.* Chicago: University of Chicago Press.

McNeill, D. (2005). *Gesture and thought.* Chicago: University of Chicago Press.

Merleau-Ponty, M. (1962). *Phenomenology of perception* (C. Smith, Trans.). London: Routledge.

Mesulam, M.-M. (1990). Large-scale neurocognitive networks and distributed accessing for attention, language, and memory. *Annals of Neurology, 28*, 597–613.

Milner, A. D., & Goodale, M. A. (2008). Two visual systems re-viewed. *Neuropsychologia, 46*(3), 774–785. doi: 10.1016/j.neuropsychologia.2007.10.005.

Moczek, A. P., Sears, K. E., Stollewerk, A., Wittkopp, P. J., Diggle, P., Dworkin, I., . . . Extavour, C. G. (2015). The significance and scope of evolutionary developmental biology: A vision for the twenty-first century. *Evolution & Development, 17*(3), 198–219. doi: 10.1111/ede.12125.

Neafsey, E. J. (1990). Prefrontal cortical control of the autonomic nervous system: Anatomical and physiological observations. In H. B. M. Uylings, C. G. Van Eden, J. P. C. De Bruin, M. A. Corner, & M. G. P. Feenstra (Eds.), *The prefrontal cortex: Its structure, function and pathology* (pp. 147–166). New York: Elsevier.

Neafsey, E. J., Terreberry, R. R., Hurley, K. M., Ruit, K. G., & Frysztak, R. J. (1993). Anterior cingulate cortex in rodents: Connections, visceral control functions, and implications for emotion. In B. A. Vogt & M. Gabriel (Eds.), *Neurobiology of the cingulate cortex and limbic thalamus* (pp. 206–223). Boston: Birkhauser.

Niedermeyer, E. (1966). Generalized seizure discharges and possible precipitating mechanisms. *Epilepsia, 7*(1), 23–29. Retrieved from https://pubmed.ncbi.nlm.nih.gov/5222470.

Niedermeyer, E. (2003). Electrophysiology of the frontal lobe. *Clinical Electroencephalography, 34*(1), 5–12. Retrieved from https://pubmed.ncbi.nlm.nih.gov/12515445.

Niewenhuys, R., Ten Donkelaar, H., & Nicholson, C. (1998). *The central nervous system of vertebrates* (3 vols.). Berlin: Springer.

Nussbaum, M. C. (2001). *The fragility of goodness: Luck and ethics in Greek tragedy and philosophy*. Cambridge: Cambridge University Press.

Obrig, H., Neufang, M., Wenzel, R., Kohl, M., Steinbrink, J., Einhaupl, K., & Villringer, A. (2000). Spontaneous low frequency oscillations of cerebral hemodynamics and metabolism in human adults. *Neuroimage, 12*(6), 623–639. Retrieved from https://pubmed.ncbi.nlm.nih.gov/11112395.

Pappas, G. (2016). John Dewey's radical logic: The function of the qualitative in thinking. *Transactions of the Charles S. Peirce Society, 52*(3), 435–468.

Pearlson, G. D., & Robinson, R. G. (1981). Suction lesions of the frontal cerebral cortex in the rat induce asymmetrical behavioral and catecholaminergic responses. *Brain Research, 218*, 233–242.

Peirce, C. S. (1931). *Collected papers of Charles Sanders Peirce*. Cambridge, MA: Harvard University Press.

Pfeifer, J. H., & Allen, N. B. (2012). Arrested development? Reconsidering dual-systems models of brain function in adolescence and disorders. *Trends in Cognitive Sciences, 16*(6), 322–329. doi: 10.1016/j.tics.2012.04.011.

Piaget, J. (1936/1992). *The origins of intelligence in children*. New York: International Universities Press.

Piaget, J. (1971). *Genetic epistemology*. New York: W. W. Norton.

Polanyi, M. (1966). The logic of tacit inference. *Philosophy, 41*(155), 1–18.

Popper, K. (1959). *The logic of scientific discovery*. New York: Routledge Classics.

Posner, M. I., & Dehaene, S. (1994). Attentional networks. *Trends in Neurosciences, 17*(2), 75–79.

Posner, M. I., & Keele, S. W. (1968). On the genesis of abstract ideas. *Journal of Experimental Psychology, 77*(3), 353–363. Retrieved from https://pubmed.ncbi.nlm.nih.gov/5665566.

Posner, M. I., & Raichle, M. E. (1994). *Images of mind*. New York: Scientific American Library.

Posner, M. I., & Rothbart, M. K. (2000). Developing mechanisms of self-regulation. *Development and Psychopathology, 12*(3), 427–441. Retrieved from https://pubmed.ncbi.nlm.nih.gov/11014746.

Posner, M. I., & Rothbart, M. K. (2009). Toward a physical basis of attention and self regulation. *Physics of Life Reviews, 6*(2), 103–120. Retrieved from https://www.ncbi.nlm.nih.gov/pmc/articles/PMC2748943.

Pribram, K. H., & Gill, M. M. (1976). *Freud's "project" re-assessed*. New York: Basic Books.

Quine, W. V. O. (1951). Two dogmas of empiricism. *Philosophical Review, 60,* 20–43.

Quine, W. V. O. (1960). *Word and object.* Cambridge, MA: MIT Press.

Raichle, M. E., & Gusnard, D. A. (2005). Intrinsic brain activity sets the stage for expression of motivated behavior. *Journal of Comparative Neurology, 493*(1), 167–176. Retrieved from https://pubmed.ncbi.nlm.nih.gov/16254998.

Reddy, M. (1979). The conduit metaphor. In A. Ortony (Ed.), *Metaphor and thought* (pp. 283–324). Cambridge: Cambridge University Press.

Robinson, S., & Pallasmaa, J. (2015). *Mind in architecture: Neuroscience, embodiment, and the future of design.* Cambridge, MA: MIT Press.

Rorty, R. (1967). *The linguistic turn.* Chicago: University of Chicago Press.

Rorty, R. (1979). *Philosophy and the mirror of nature.* Princeton, NJ: Princeton University Press.

Rorty, R. (1989). *Contingency, irony, and solidarity.* Cambridge: Cambridge University Press.

Rosch, E., Mervis, C. B., Gray, W. D., Johnson, D. M., & Boyes-Braem, P. (1976). Basic objects in natural categories. *Cognitive Psychology, 8*(1), 382–439.

Rumelhart, D. E., & McClelland, J. L. (1986). *Parallel distributed processing: Explorations in the microstructure of cognition: Vol. 1. Foundations.* Cambridge, MA: MIT Press.

Schulkin, J. (2011). *Adaptation and well-being.* Cambridge: Cambridge University Press.

Schulkin, J., & Sterling, P. (2019). Allostasis: A brain-centered, predictive mode of physiological regulation. *Trends in Neurosciences, 42*(10), 740–752.

Seeley, W. W., Menon, V., Schatzberg, A. F., Keller, J., Glover, G. H., Kenna, H., . . . Greicius, M. D. (2007). Dissociable intrinsic connectivity networks for salience processing and executive control. *Journal of Neuroscience, 27*(9), 2349–2356. doi: 10.1523/JNEUROSCI.5587-06.2007.

Semendeferi, K., Armstrong, E., Schleicher, A., Zilles, K., & Van Hoesen, G. W. (2001). Prefrontal cortex in humans and apes: A comparative study of area 10. *American Journal of Physical Anthropology, 114*(3), 224–241. doi: 10.1002/1096-8644 (200103)114:3<224::AID-AJPA1022>30.CO;2-I.

Shannon, C. E. & Weaver, W. (1949). *The mathematical theory of communication.* Champaign: University of Illinois Press.

Shapiro, D. (1965). *Neurotic styles.* New York: Basic Books.

Shaw, E. D., Mann, J. J., Stokes, P. E., & Menvitz, Z. A. (1986). Effects of lithium carbonate on associative productivity and idiosyncrasy in bipolar outpatients. *American Journal of Psychiatry, 143,* 1166–1169.

Sheets-Johnstone, M. (1999). *The primacy of movement.* Amsterdam: John Benjamins.

Shepard, R. N. (1984). Ecological constraints on internal representation: Resonant kinematics of perceiving, imagining, thinking, and dreaming. *Psychological Review, 91*(4), 417–447.

Shima, K., & Tanji, J. (2000). Neuronal activity in the supplementary and presupplementary motor areas for temporal organization of multiple movements. *Journal of Neurophysiology, 84*(4), 2148–2160. Retrieved from https://pubmed.ncbi.nlm.nih.gov/11024102.

Shimamura, A. P., & Squire, L. R. (1986). Memory and metamemory: A study of the feeling-of-knowing phenomenon in amnesic patients. *Journal of Experimental Psychology: Learning, Memory, and Cognition, 12*(3), 452–460.

Slingerland, E. (2008). *What science offers the humanities: Integrating body and culture.* Cambridge: Cambridge University Press.

Snow, C. P. (1959). *The two cultures.* London: Cambridge University Press.

Solymosi, T., & Shook, J. (2014). *Neuroscience, neurophilosophy and pragmatism: Brains at work with the world.* New York: Palgrave Macmillan.

Squire, L. R., & Spanis, C. W. (1984). Long gradient of retrograde amnesia in mice: Continuity with the findings in humans. *Behavioral Neuroscience, 98*(2), 345–348.

Squire, L. R., Wetzel, C. D., & Slater, P. C. (1978). Anterograde amnesia following ECT: An analysis of the beneficial effects of partial information. *Neuropsychologia, 16*(3), 339–348.

Squire, L. R., & Zola, S. M. (1997). Amnesia, memory and brain systems. *Philosophical Transactions of the Royal Society of London. Series B: Biological Sciences, 352*(1362), 1663–1673.

Squire, L. R., Zola-Morgan, S., Cave, C. B., Haist, F., Musen, G., & Suzuki, W. A. (1990). Memory: Organization of brain systems and cognition. *Cold Spring Harbor Symposia on Quantitative Biology, 55*, 1007–1023.

Tononi, G. (2008). Consciousness as integrated information: A provisional manifesto. *Biological Bulletin, 215*(3), 216–242. doi: 10.2307/25470707.

Tononi, G., & Edelman, G. M. (1998). Consciousness and complexity. *Science, 282*, 1846–1851.

Tononi, G., & Koch, C. (2008). The neural correlates of consciousness: An update. *Annals of the New York Academy of Sciences, 1124*, 239–261. doi: 10.1196/annals.1440.004.

Trevarthen, C., & Aitken, K. J. (2001). Infant intersubjectivity: Research, theory, and clinical applications. *Journal of Child Psychology and Psychiatry, 42*(1), 3–48. Retrieved from https://pubmed.ncbi.nlm.nih.gov/11205623.

Tucker, D. M. (1991). Developing emotions and cortical networks. In M. R. Gunnar & C. A. Nelson (Eds.), *The Minnesota symposia on child psychology* (Vol. 24, pp. 75–128). Hillsdale, NJ: Lawrence Erlbaum.

Tucker, D. M. (2002). Embodied meaning: An evolutionary-developmental analysis of adaptive semantics. In T. Givon & B. Malle (Eds.), *The evolution of language out of pre-language* (pp. 51–82). Amsterdam: John Benjamins.

Tucker, D. M. (2007). *Mind from body: Experience from neural structure.* New York: Oxford University Press.

Tucker, D. M. (Unpublished). *Turning left and right: The unconscious foundations of political thought.*

Tucker, D. M., Brown, M., Luu, P., & Holmes, M. D. (2007). Discharges in ventromedial frontal cortex during absence spells. *Epilepsy and Behavior, 11,* 546–557. Retrieved from https://pubmed.ncbi.nlm.nih.gov/17728188.

Tucker, D. M., & Derryberry, D. (1992). Motivated attention: Anxiety and the frontal executive functions. *Neuropsychiatry, Neuropsychology, and Behavioral Neurology, 5,* 233–252.

Tucker, D. M., Derryberry, D., & Luu, P. (2000). Anatomy and physiology of human emotion: Vertical integration of brainstem, limbic, and cortical systems. In J. Borod (Ed.), *Handbook of the neuropsychology of emotion* (pp. 56–79). New York: Oxford University Press.

Tucker, D. M., & Holmes, M. D. (2010). Fractures and bindings of consciousness. *American Scientist, 99,* 32–39.

Tucker, D. M., & Luu, P. (2006). Adaptive binding. In H. Zimmer, A. Mecklinger, & U. Lindenberger (Eds.), *Binding in human memory: A neurocognitive approach* (pp. 85–108). New York: Oxford University Press.

Tucker, D. M., & Luu, P. (2007). Neurophysiology of motivated learning: Adaptive mechanisms of cognitive bias in depression. *Cognitive Therapy and Research, 31,* 189–209.

Tucker, D. M., & Luu, P. (2012). *Cognition and neural development.* New York: Oxford University Press.

Tucker, D. M., Luu, P., Desmond, R. E., Jr., Hartry-Speiser, A., Davey, C., & Flaisch, T. (2003). Corticolimbic mechanisms in emotional decisions. *Emotion, 3*(2), 127–149. Retrieved from https://pubmed.ncbi.nlm.nih.gov/12899415.

Tucker, D. M., Luu, P., & Pribram, K. H. (1995). Social and emotional self-regulation. *Annals of the New York Academy of Sciences, 769,* 213–239.

Tucker, D. M., Poulsen, C., & Luu, P. (2015). Critical periods for the neurodevelopmental processes of externalizing and internalizing. *Development and Psychopathology, 27*(2), 321–346. doi: 10.1017/S0954579415000024.

Tucker, D. M., & Williamson, P. A. (1984). Asymmetric neural control systems in human self-regulation. *Psychological Review, 91*(2), 185–215.

Turner, M. (1991). *Reading minds: The study of English in the age of cognitive science.* Princeton, NJ: Princeton University Press.

Ungerleider, L. G., & Mishkin, M. (1982). Two cortical visual systems. In D. J. Ingle, R. J. W. Mansfield, & M. A. Goodale (Eds.), *The analysis of visual behavior* (pp. 549–586). Cambridge, MA: MIT Press.

Varela, F. J., Thompson, E. T., & Rosch, E. (1991). *The embodied mind: Cognitive science and human experience.* Cambridge, MA: MIT Press.

von de Malsburg, C., & Singer, W. (1988). Principles of cortical network organization. In P. Rakic & W. Singer (Eds.), *Neurobiology of neocortex* (pp. 69–99). New York: Wiley.

Walker, M. P. (2008). Sleep-dependent memory processing. *Harvard Review of Psychiatry, 16*(5), 287–298. doi: 10.1080/10673220802432517.

Walker, M. P. (2009). The role of sleep in cognition and emotion. *Annals of the New York Academy of Sciences, 1156,* 168–197. doi: 10.1111/j.1749-6632.2009.04416.x.

Wehrs, D. R., & Blake, T. (2017). *The Palgrave handbook of affect studies and textual criticism.* Cham, Switzerland: Springer.

Wei, Y., Krishnan, G. P., Komarov, M., & Bazhenov, M. (2018). Differential roles of sleep spindles and sleep slow oscillations in memory consolidation. *PLoS Computational Biology, 14*(7), e1006322.

Werner, H. (1957). *The comparative psychology of mental development.* New York: Harper.

Whitehead, A. N. (1925/1997). *Science and the modern world.* New York: Free Press.

Whitehead, A. N. (1929/2010). *Process and reality.* New York: Simon and Schuster.

Whitehead, A. N. (1933). *Adventures of ideas.* New York: Free Press.

Whitehead, A. N. (1938). *Modes of thought.* New York: Free Press.

Wiener, N. (1961). *Cybernetics, or control and communication in the animal and the machine.* Cambridge, MA: MIT Press.

Wiley, N. (1994). *The semiotic self.* Chicago: University of Chicago Press.

Yonelinas, A. (2006). Unpacking explicit memory: Separating recollection and familiarity. In H. D. Zimmer, A. Mecklinger, & U. Lindenberger (Eds.), *Handbook of binding and memory: Perspectives from cognitive neuroscience.* Saarbrücken, Germany: Oxford University Press.

Index

f